MW01533698

PREVENTING FOREIGN MATERIAL CONTAMINATION OF FOODS

PREVENTING FOREIGN MATERIAL CONTAMINATION OF FOODS

Doug Peariso

Doug Peariso is the former senior quality assurance manager for the Foods Business Unit of Gerber Products Company, Fremont, Michigan, where he was accountable for the oversight and strategic leadership of the quality systems for Gerber's Domestic Baby Food Operations. Previously Peariso held quality assurance posts with The Clorox Company (Parsons, West Virginia) and Campbell Soup Company/Vlasic Foods International (Imlay City, Michigan).

©2006 Blackwell Publishing
All rights reserved

Blackwell Publishing Professional
2121 State Avenue, Ames, Iowa 50014, USA

Orders: 1-800-862-6657
Office: 1-515-292-0140
Fax: 1-515-292-3348
Web site: www.blackwellprofessional.com

Blackwell Publishing Ltd
9600 Garsington Road, Oxford OX4 2DQ, UK
Tel.: +44 (0)1865 776868

Blackwell Publishing Asia
550 Swanston Street, Carlton, Victoria 3053, Australia
Tel.: +61 (0)3 8359 1011

Authorization to photocopy items for internal or personal use, or the internal or personal use of specific clients, is granted by Blackwell Publishing, provided that the base fee of $.10 per copy is paid directly to the Copyright Clearance Center, 222 Rosewood Drive, Danvers, MA 01923. For those organizations that have been granted a photocopy license by CCC, a separate system of payments has been arranged. The fee code for users of the Transactional Reporting Service is 0-8138-1639-4/2006 $.10.

First edition, 2006

Library of Congress Cataloging-in-Publication Data

Peariso, Doug.
Preventing foreign material contamination of foods / Doug Peariso.—1st ed.
p. cm.
Includes bibliographical references and index.
ISBN 0-8138-1639-4 (alk. paper)
1. Food industry and trade—Quality control. 2. Food contamination—Prevention. I. Title.

TP372.5.P35 2005
664'.07—dc22

2004023184

The last digit is the print number: 9 8 7 6 5 4 3 2 1

Dedication

This text is dedicated to the memory of Janette M. Forrest. Her unrelenting faith and tenacity to push through the difficult periods in life were exemplary traits. These traits were only surpassed by the strength and courage that she modeled during the arduous final months of her life. The living example of character that she portrayed during this time will forever serve as an inspiration, and a benchmark, for how I will live the rest of my life. She will continue to be missed immensely by her family and friends.

Contents

PREVENTING FOREIGN MATERIAL CONTAMINATION OF FOODS

CHAPTER 1

The Importance of Preventing Foreign Material Contamination of Food Products

1.1 Introduction

The purpose of this chapter is to introduce several concepts that will serve as the conceptual basis for the rest of the text. It is critical that the reader of this text have an appreciation not only for the potential sources, definitions, and control mechanisms for foreign materials in food products, but also a comprehension of what can be gained by preventing their entrance into the food supply. A general discussion centered on the evolution of the food industry, relevant trends in the consumption patterns of food products, and the expectations of consumers will be initially detailed. Progression through the chapter should, hopefully, make the reader more aware of the benefits to be gained via application of foreign material control programs.

1.2 The ever-changing role of the food processor from previous eras to the present

In 1906 no consumer would complain about finding a single stem in a can of green beans, let alone a stray peanut in a chocolate bar. "Macerate and move on" was essentially the credo of the era. For those who may not relate to this statement, a quick excerpt from Upton Sinclair's (18) classic text, *The Jungle,* should afford a quick calibration of the prevalent manufacturing practices of the era:

> This is no fairy story and no joke; the meat will be shoveled into carts and the man who did the shoveling will not trouble to lift out a rat even when he saw one.
>
> –*The Jungle*–Chapter 14, pg. 162

After reading such great prose, it is easy to understand why the general public cried out for government oversight of the manufacturers of food products.

During the first decades of the 20th century, most food products were sourced locally and consumed within their respective seasons. Deseasonalization of food consumption patterns was aided by the proliferation of such practices as expansion of the cold chain, increased prevalence of truck farming, and an increase in the number of thermally processed food products (11). Developments in canning and ever-increasing capacity of food manufacturers also provided a crucial platform to further aid society in the deseasonalization of food consumption patterns. As these practices matured consumers were no longer tithed to the harvest moon, the root cellar, or the barnyard slaughter stall to provide nourishment for their families. Seasonal commodities were now being transformed into value-added products. Retailers, restauranteers, and processors alike started to cash in on this new paradigm and provide these products to an eager public.

The present state of the food industry is quite different from the early era alluded to in the opening paragraph of this chapter. Growers, herders, processors, and foodservice locations alike have experienced vast changes since this period, and more complex business models have been adopted to fill new niches within the consumer market. These niches were created via the expansion of the population known as the middle class. The citizens of today's middle class have not only increased in number, but possess substantially greater disposable income than the previously existing regime. These consumers are spending an ever-increasing amount of monies on food products. Food-based retailing accounted for 23 percent of the $3.78 trillion retail trade within the United States in 2003 (10).

This "new" middle class represents an eager consumer base toward which food companies can actively market new and existing products. The margin contributions of a modern food company are now no longer justified by merely providing a value-added agricultural product to consumers. Consumers today no longer view having access to seasonal products during nonseasonal periods as nouveau. Products that were previously seen as value-added, such as canned foods, ground meats, and ready-to-eat (RTE) products are now viewed as commodities themselves. Presently the critical aspects of what consumer-perceivable value is added to these new commodities, and how efficiently this is done by the processor, is largely what justifies the long-term operating margin and market share potential of a company. A key "what" of today's food products required to command pricing premiums are innovation and convenience factors.

Bustling lifestyles, decreased contact times, and nonnuclear families not only make convenience essential for the consumer of the "hot and now" era, but also make convenience products command a premium price at the point of sale. An excellent marketing strategy can also go a long way to impart short-term value-added perceptions about products upon consumers who purchase products. Excellent marketing campaigns can even invoke sudden renewed interest in existing products and deliver increased sales for a period of time. Marketing is merely positioning of the product, however, and long-term success of a brand will still be driven by the "what" and "how" factors defined above.

Another key difference found in the present era is that of novel information platforms. Today's "hot and now" consumer is continuously fed misinformation about the food industry via propagation of falsehoods and urban legends on the Internet. The Internet also offers new fertile ground for special interest groups to rapidly spawn propaganda that was created by people of varying credibility and knowledge. The ever-increasing and far-reaching coverage afforded by the mass media also represents a challenge to food companies. Coverage of negative events (such as recalls and other significant public health events like tampering) tend to be drawn out and subject to rebroadcast as news of interest. The modern food industry professional must take this propaganda machine into account when making business decisions related to operational practices or disposition of food products.

1.3 Definitions of foreign material

It is important that a definition of foreign material (FM) is presented before progressing further with the text. FM, referred to in various circles as foreign objects and/or extraneous matter, is defined as any material(s) whose quantity and presence within a food product is unanticipated by: (1) the customer or consumer purchasing the product, and/or (2) the regulatory body that has jurisdiction over the processing and distribution of the product.

This definition is slightly different from previous renditions, but hopefully it is a better model for the consumer and regulatory expectations for products. The potential number of situations that could be categorized within these two subdivisions represent a myriad of possibilities. It is appropriate at this time to now introduce the current definitions and references made to FM from the position of the regulatory agencies. The regulatory definitions are narrower in scope than the definition above, but they must be well understood by the processor in order to ensure that their ongoing decision-making process is based on a solid knowledge base.

The United States Food and Drug Administration (FDA) has adopted a broader and differing definition of foreign materials that is quite definitive in scope. Within the Food, Drug, and Cosmetic Act there are three citations that are relevant in defining foreign materials from Section 402(a) (14 and 18):

1. "402(a)(1), details the position that "a food shall be deemed to be adulterated if it bears or contains any poisonous or deleterious substance which may render it injurious to health."
2. "402(a)(3), states "a food shall be deemed to be adulterated if it consists in whole or in part of any filthy, putrid, or decomposed substance, or if it is otherwise unfit for food."
3. "402(a)(4), states "a food shall be deemed to be adulterated if it has been prepared, packed, or held under insanitary conditions whereby it may have become contaminated with filth, or whereby it may have been rendered injurious to health."

The FDA also references foreign materials within their published Good Manufacturing Practices (GMP). Section 110.80(b)(8) of the GMP states that:

> Effective measures shall be taken to protect against the inclusion of metal or other extraneous materials in food. Compliance with this requirement may be accomplished by using sieves, traps, magnets, electronic metal detectors, or other suitable effective means.

The United States Department of Agriculture (USDA) Food Safety and Inspecton Service (FSIS) has recently redefined FM via issuance of FSIS

Directive 7310.5. Within Section V (terminology) of this directive, FM is referenced to and defined as follows:

> Foreign material: Foreign materials are non-animal objects, such as metal, plastic, rubber, glass, wood, steel, or lead shot.

FSIS, like the FDA, also acknowledges the presence of such materials as adulterants.

1.3.1 Classification of foreign material contaminants and risk assessment

A clear distinction must be made early in this text that FM must be first separated into one of three categories before applying management strategies or controls. The three categories revolve around the key question of whether or not the FM represents a:

1. Threat of significance to public health
2. Potential regulatory concern
3. Consumer-perceivable aesthetic issue (these can sometimes still represent a regulatory concern)

It is interesting to note that some foreign material contaminants may many times fall into more than one category (at times existing in all three categories simultaneously). Each category will be discussed in the pages that follow, but realize that the presence of FM within a food product, regardless of its respective class, should be handled delicately by the processor.

1.3.1.1 FM that represents a public health concern

For FM to be termed as a public health concern, it must represent a biological, chemical, or physical hazard to the end user of the product through either its inclusion in, or its interaction with, the product. Both the FSIS and the FDA maintain a set of internal standards that aid them in defining foreign materials as physical hazards. FDA defines the criteria necessary for a food product to be considered a physical hazard in Chapter 5 of its ORA compliance policy guide—specifically Subchapter 555 issued on March 23, 1999. The regulatory action guidance stated by FDA is as follows:

> The following represent the criteria for direct reference seizure to the Division of Compliance Management and Operations (HFC-210) and direct reference import detention to the districts:
>
> a. The product contains a hard or sharp foreign object that measures 7 mm to 25 mm in length.
>
> and
>
> b. The product is ready to eat, or according to instructions or other guidance or requirements, it requires only minimum preparation steps; e.g., heating, that would not eliminate, invalidate, or neutralize the hazard prior to consumption.
>
> Samples found to contain foreign objects that meet criteria a. and b., above, should be considered adulterated within the meaning of 21 U.S.C. 342(a)(1).

The policy further expounds upon how the presence of FM within a food product that doesn't meet the above criteria may still require further review by USDA's Center for Food Safety and Applied Nutrition (CFSAN) to dispel it as a hazard to public health. The FDA also has communicated that hard or sharp foreign objects between 7 mm and 2 mm may still represent a health hazard to certain subpopulations (namely infants and elderly). The hazard potential for these hard or sharp foreign objects in the food product is well defined in the current body of scientific knowledge. The health hazard potential of these objects resides in the areas of potential laceration, dental injury, other gastrointestinal trauma (perforations), and/or sources of infection. The rationale for the classification of FM as physical hazards is excellently presented in publications prepared by A.R. Olsen and others (12, 13, and 14).

The FSIS has also defined criteria for identifying FM as potentially hazardous to the public. The topic of FM inclusions in food products was thoroughly discussed at a two-day public conference held by the agency in Omaha, Nebraska, on September 24–25, 2002. Representatives from food processing firms, academia, and the USDA openly discussed what criteria would dictate that an FM inclusion be defined as a hazard to public health. The agency stated that the FDA standard was valid, and scientifically justifiable. The officials present expressed that FSIS follows similar

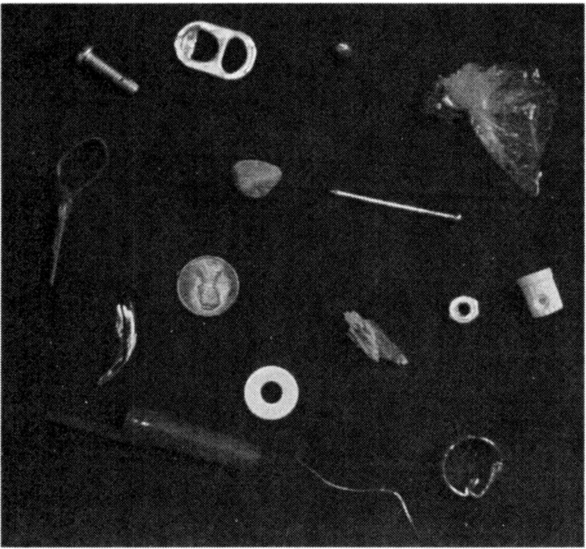

Figure 1.1. Examples of foreign materials that meet the size and hardness criteria to be considered physical hazards in food products.

size criteria (7 to 25 mm) in determining public health hazards presented by hard or sharp FM. FSIS also alluded to the fact that they will evaluate complaints and sampling finds on a case-by-case basis (just like FDA) even if the size and preparation criteria are not met. Figure 1.1 illustrates examples of hard and sharp contaminants commonly generating complaints in the food industry. In the Omaha review, the FSIS also made reference to choking hazards presented by foreign materials, and the relevant standards used by the United States Consumer Products Safety Commission (CPSC).

The CPSC maintains standards to classify objects as choking hazards to children under three years of age. Essentially, the discussion at the FSIS conference centered on the fact that the use of a small-parts gauge (a cylinder of specific dimensions that is supposed to mimic the airway of a child) has aided in classifying objects as choking hazards for children. The CPSC provides guidance in 16 CFR 1501.4 Parts (a) and (b) that detail the methodology used to determine if a detached object represents a choking hazard for children under three years of age. Figure 1.2 displays the illustration from the U.S. Code of Federal Regulations (CFR) for the fixture that is to be used to test the small parts.

Consumer Product Safety Commission § 1501.5

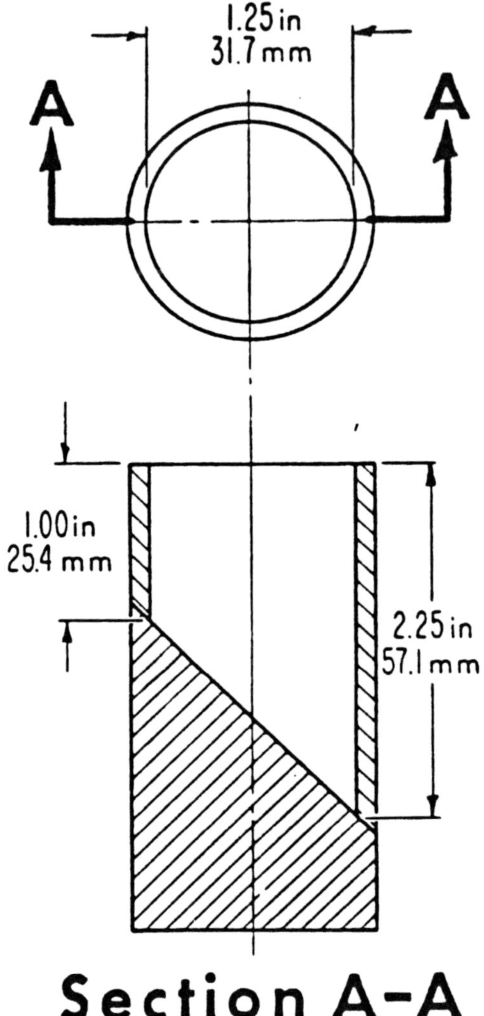

FIG 1-SMALL PARTS CYLINDER

Figure 1.2. Diagram of a small parts cylinder used to determine if a product represents a choking hazard to children.

1.3.1.1.1 Foreign material can be more than a physical hazard

The concept that FM may represent more than a physical hazard is generally well understood by food safety and quality practitioners. The contemporary logic used by most personnel in manufacturing facilities may differ, though. This group tends to define foreign materials as either hazardous to public health or nonhazardous to public health by screening the object through a single dimension—"Does it meet the definition of a physical hazard?" Figure 1.3 provides an illustration of this flawed logic.

There are two other dimensions that must be considered when classifying foreign material contaminants as public health hazards. The first dimension is related to the question, "Does the material itself represent a potential biological or chemical hazard of significance to public health?" When the correct resources are participating in the discussion, the direct

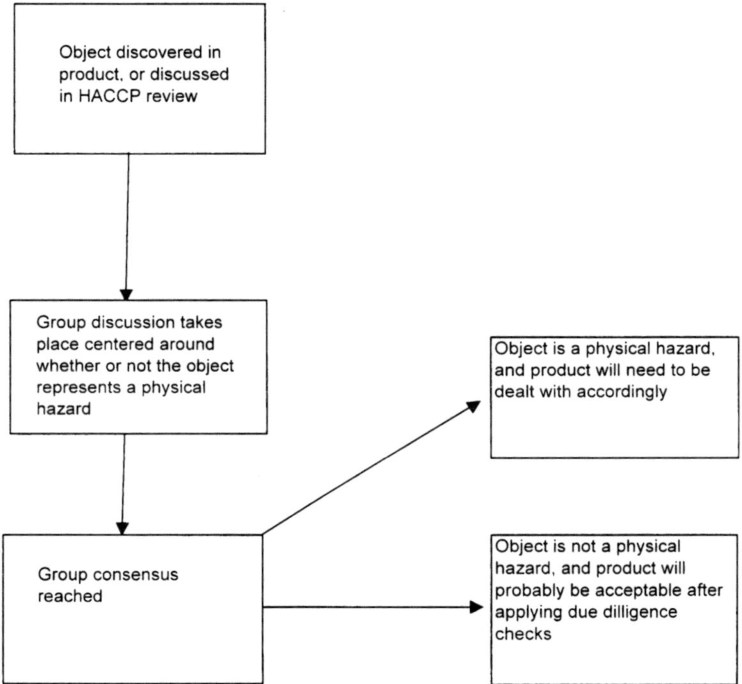

Figure 1.3. Flawed one-dimensional logic of classifying all foreign materials as either physical hazards or being incidental contaminants possessing limited risk to public health.

relationship of the material to other hazard classifications is usually an easy question to answer.

The second dimension of analysis required to determine the potential public health implications that FM inclusions within a product may represent is an interaction effect. The interaction scenario can be defined by analyzing whether the foreign material would nullify the effectiveness of prerequisite programs or critical control points required to deliver a safe product to the consumer. This determination needs to be 100 percent correlative with the contemporary body of knowledge available.

A great example of a failure to determine whether or not an object represents a public health hazard would be the unanticipated loss of a lid from a plastic ingredient pail. Such containers are commonly used in the blending areas of beverage manufacturing facilities, and they sometimes disappear. When this happens, there is great potential for thousands of gallons of beverages to be produced before a routine monitoring check discovers the ground-up plastic pieces present in the process stream. Typical discussions following a contamination event such as this one usually center on the plastic material introduced, and what it means for the product disposition. Such a discussion may include the fact that the physical filtration of the product via a downstream control (strainer, filter bag, etc.) would assure that no plastic would be evident in the finished product. This, in most cases, is a true statement. Unfortunately, ending the discussion under this single-dimension model presents the opportunity for group think to conclude that the finished product would still be fine, and no extensive product hold is warranted. Without infusion of multidimensional logic into such a discussion, personnel present may forget to specifically question the origin of the lid, or what ingredients it was in contact with previously. This is unfortunate because, as an example, the pail that the lid originated from in this case had contained a dairy ingredient used during the previous production day, but the product that was being produced when the plastic was found did not contain any dairy ingredients. The plastic lid fragments in this case actually represent a vector for an allergen—a chemical hazard. Other examples of this type of direct hazard potential are the presence of wheat kernels in a batch of nonwheat-containing flour, and the presence of imbedded dirt or grease (potential biological or chemical hazard) on packaging materials (films) used for production of ready-to-eat products.

The interaction scenarios defined above usually are not as easy to distinguish. Stray pieces of wood (originating from a pallet) found present within a pouch of thermally processed low- acid product (soup stock, nutritional beverage, etc.) is most definitely an FM issue. In practice, a number of disposition strategies may be envisioned initially by the processor in this situation, but how often would the interaction effects be questioned? The interaction scenario in this case is that the microorganisms that may be present within the wood could be given a safe harbor from the thermal process applied, or the particle may disrupt conduction of heat through the product. Either way this would render the commercial sterility of the food product questionable, and the event would need to be treated as a process deviation to be reviewed by a thermal process authority, not a statistical lot inspection activity to find more stray chunks of wood.

Another interesting example of the interaction scenario can be found in the vegetable canning world. The presence of stems in products is a common issue for processors of vegetable and fruit products. Products containing numerous stray stems, that may not even represent a consumer-perceivable aesthetic issue, could protrude over the edge of a container and nest in the area of the final hermetic seal. The area that the stem is occupying could very well leak at a slow rate, allowing the container to pass through common detection technologies (i.e., dud detectors, inversion, squeeze tests, fill-level detection, etc.) at the facility before the breach is discernable. Post-process microbial contamination is likely in this scenario.

The examples presented above are usually not considered reasonably likely to occur in most food processing establishments, and many times the systems in place at these facilities justify this position. When dealing with the many unforeseen or "unique" situations that tend to present themselves throughout one's career, interaction scenarios are going to be evident. Hopefully the examples presented above help to provide credence to the concept that evaluation of the risks associated with an FM contaminant must be multidimensional and go beyond the aspects of the physical hazard.

1.3.1.2 Regulatory compliance and unavoidable contaminants

Public health protection is paramount for the food industry; however, maintaining regulatory compliance may become difficult when dealing with common foreign materials. A substance may be present in a product that is generally regarded as safe (GRAS) by regulatory agencies, and only visible with a 20× optical microscope. Such an inclusion would not be discernable even by the most finicky consumer. The problem in this situation is the fact that, unless this foreign substance is defined as an unavoidable contaminant and granted clemency by published defect action levels or other regulatory agency directives, its presence is not legally allowable. The product can be considered by both the FDA and FSIS as adulterated or otherwise mislabeled due to inclusion of filth, impurities, or nonlabeled ingredients. Figure 1.4 displays a common example of such a contaminant. Microscopic metal (in this case cast iron) is a common issue in the food industry, and the processor must proceed cautiously when dealing with its presence.

Additionally, if your products contain the exact foreign materials allowed under published guidelines or standards of identity (such as CFSAN's Defect Action Levels [DALs]), there are thresholds that cannot be exceeded. In most cases these thresholds should not even be approached by the processor with any regularity because they are based on the minimum purity level that modern technology can achieve at a minimum financial burden. A perfect example of this type of regulatory issue is an applesauce producer who ran into insect fragment issues. If too many fragments are found to be present within a lot sample, then the purity of the product or the conditions that it was manufactured or held under could be defined as unsatisfactory by the regulatory agency. The product could then be considered adulterated, and be subject to seizure, recall, or other regulatory actions.

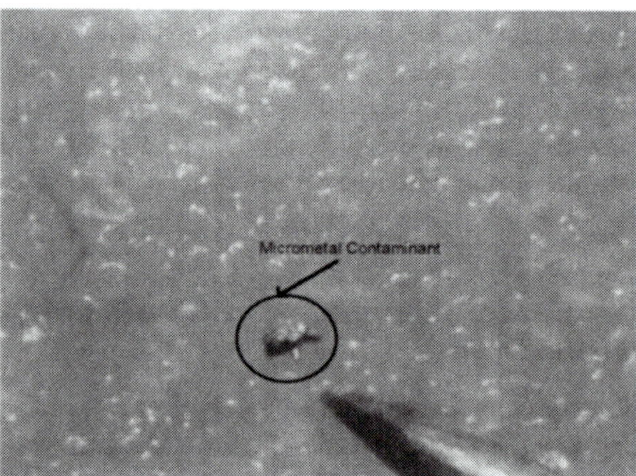

Figure 1.4. Microscopic metal contamination present at 60 × magnification in a common foodservice condiment.

1.3.1.3 The needs of the customer or consumer

The final example of foreign material classification is product quality specifications, or consumer-perceivable aesthetic concerns. If the presence of a related (or unrelated) FM in your product does not represent a threat to public health and is technically allowable by the regulatory agencies that govern the product, it is not an issue, right? Unfortunately this logic may not be true. One must still understand whether or not the situation presents a risk of customer/consumer dissatisfaction. A classic example of customer dissatisfaction can be found when examining a processor's relationship with a foodservice customer, such as a fast-food chain. If the chain purchases 3.5 L containers of sliced banana peppers destined to be placed on a sub sandwich, then they usually express an expectation about the number of good slices that should be present. The customer may not desire more than ten seed cavities, or six stems per 500 g of peppers. Likewise, a consumer may not appreciate a lengthy piece of tendon being present within a ready-to-eat ground and formed chicken product. The situations above do not readily present a risk to public health, nor do they reflect a regulatory concern, but they could jeopardize the relationship and repeat purchase intent of the customer or consumer.

1.3.1.4 Documenting the rationale for placing FM within a category

Hopefully it is apparent to the reader that, whenever an FM issue is encountered, the utmost due diligence is required by the processor. All investigation of ruling out a food safety risk should be documented for future consultation if an incident or investigation were to occur. A good general screening guide for your problem can be obtained by utilizing the classification guide represented in Table 1.1.

1.4 Defining the current scope of the issue in the marketplace—consumer complaints and recalls

Consumer complaint data is extremely privileged information that is almost never made publicly available by food processors. The FSIS and FDA, however, regularly receive complaints from consumers in an unsolicited manner. The agencies occasionally publish summaries of these

Table 1.1 Key questions for placement of foreign materials into 3 broad categories.

Public Health Yes/No

1. Could the object/material serve as a vector for known Foodbourne pathogens, or interfere with further processing that delivers a log reduction target for pathogens of public health concern that may be inherent to the product?
2. Is the object/material in question known to exhibit deleterious toxic effects in humans and animals?
3. Is the object/material in question something that further processing will render harmful to the end consumer?
4. Is the object/material able to evoke an allergic reaction in sensitized consumers?
5. Could the object cause choking or injury to the consumer if it were to be consumed, or otherwise interfere with the digestion process?
6. Does the object violate the *dimension* rules outlined in this text for the target consumer group?
7. Could the object interfere with any hermetic sealing properties of the packaging materials?

Regulatory Yes/No

1. Is the product a naturally occurring unavoidable intrinsic contaminant as defined by CFSAN DAL Handbook, or USDA FM Directives?
2. If the answer to 1 above is yes, then is the quantity of the contaminants present within the DAL handbook's parameters when using the same procedures of analysis?
3. Is the object/material something that could not be categorized by the current ingredient statement of the finished product label panel?
4. Does the presence of the object violate any known standard identities for the product (e.g., a pit found in pitted prunes)?
5. Is it likely that the object/material would damage the food contact surface of the packaging and expose leachable nonapproved components?
6. Is the object/material something that has not been approved for direct food contact as a packaging material?

Quality or Other Legal Concerns Yes/No

1. Is the presence of this object likely to negatively impact a customer's intent for repurchase of the product?
2. Will the presence of this object violate any contractual agreements made between the company and its customer?
3. Will the presence of this object give a consumer legal standing to file a legal complaint against the company?
4. Does the presence of this object make the product less desirable than a competitor's, from the consumer's perceivable perspective?

complaints, reviewable by the general public, and these provide a unique window for viewing the scope of FM issues. FSIS presented a summary of data from their Consumer Complaint Monitoring System (CCMS) at the September 2002 conference in Omaha that helped display the scope of FM issues in the marketplace. The reporting period summarized was for calendar year 2001 through an undisclosed date in September 2002 (roughly representing 21 months of data). Within this time the agency had received 1,309 complaints, and 331 (25 percent) of these were for specific types of FM found by consumers in food products.

The complaint categories were further broken down by the type of contaminant that the consumer indicated as present within the food product. Figure 1.5 details the breakdown of the complaints as cited. Note that metal, plastic, and glass represented the majority of the complaints received while the "other" category made up a large percentage of the data (45 percent). FSIS cited that the prevalent themes in the "other" category included FM such as wood, slivers of wood, and fingernails. The agency also indicated that it had experienced a 28 percent decline in the consumer complaints it received related to extraneous materials in meat and poultry products from 1997 to 2001, and cited implementation of the prerequisite rule of Hazard Analysis Critical Control Points (HACCP) as the likely mechanism for this significant decrease in complaint contacts.

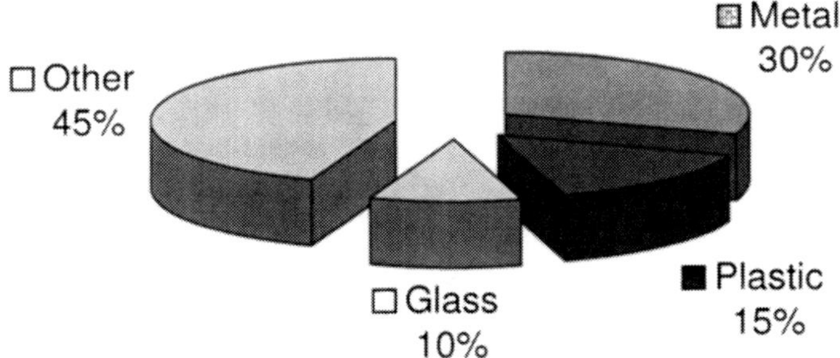

Figure 1.5. Breakdown of FM complaints from USDA's Consumer Complaint Monitoring System, January 2001–September 2002.

A second source of information that indicates the severity and scope of FM present within food products can be found through analysis of recalls conducted by food processors. Recalls are representative of the final "corrective action" available to a processor or regulatory agency for ensuring that the public is not affected by products containing FM. Analyzing the number of recalls conducted by producing firms over a period of time, with FM listed as the causal factor, can provide data on the effectiveness of contemporary approaches in dealing with that specific FM type. A personal examination of FM recalls present within the public records (available in electronic indexes maintained by FSIS and FDA, respectively) follows.

1.4.1 Summary of results from a 7.5 year retrospective analysis of food recalls conducted due to the presence of foreign materials

Recall data from January 1997 through May 2004 was compiled and evaluated for both agencies. The recall classification and FM implicated in the recall was documented and summarized. Recalls that did not indicate the type of foreign material present were excluded from the analysis. Examples of this type of situation are recalls cited by the agencies as being conducted due to possible unsanitary conditions or the presence of filth or other impurities. Recalls for allergens and the presence of insects and molds were excluded as well. Unfortunately, because the agencies do not apply use of a standard counting unit (inches, kilograms, dozens) when defining the quantity of product recalled, each situation had to be given equal weight within the analysis. This is probably not a highly negative factor in the analysis because, in the public eye, announcement of a recall is still a public recall subject to scrutiny regardless of the amount that had to be removed from distribution.

The other limitation of this data is the fact that, in most circumstances, the root cause of the FM inclusion is not made available to the public. An individual with knowledge of the industry would most likely be able to hypothesize as to the origin of the materials, based on the product that was recalled. This type of exercise was not performed for this analysis, however, due to the fact that the information would be based purely on conjecture. The facts assembled from the analysis are presented in Tables 1.2 and 1.3, and depicted graphically in Figures 1.6 and 1.7. Officials from FSIS and FDA may not completely agree with the published totals, due to

Table 1.2 Table representing results from a 7.5 year retrospective analysis of FDA recalls for foreign materials. The number of recalls by foreign material type and the relevant classifications of recalls issued is presented.

	Year	*FDA Recalls*	*Class I*	*Class II*	*Class III*	*Metal*	*Plastic/ Glass*	*Rubber*	*Other*
	1997	7	0	3	4	4	2	0	1
	1998	11	0	4	7	3	2	2	4
	1999	16	1	10	5	4	4	3	5
	2000	14	0	9	5	7	3	3	1
	2001	6	0	4	2	1	3	2	0
	2002	9	0	3	6	3	2	3	1
	2003	5	0	5	0	2	1	1	1
	2004	3	0	1	2	1	0	2	0
Total Recalls Conducted For Each Class and FM Contaminant		71	1	39	31	25	17	16	13
Percent of Total FM Recalls Represented by Each Recall Category and Contaminant			1.4	54.9	43.7	35.2	23.9	22.5	18.3

Table 1.3 Table representing results from a 7.5 year retrospective analysis of FSIS recalls for foreign materials. The number of recalls by foreign material type and the relevant classifications of recalls issued is presented.

	Year	*FDA Recalls*	*Class I*	*Class II*	*Class III*	*Metal*	*Plastic/ Glass*	*Rubber*	*Other*
	1997	6	1	5	0	2	0	4	0
	1998	6	0	6	0	4	0	0	2
	1999	2	0	2	0	0	1	1	0
	2000	5	2	3	0	2	1	1	1
	2001	6	5	0	1	1	0	5	0
	2002	5	2	2	1	1	4	0	0
	2003	3	3	0	0	1	2	0	0
	2004	4	1	2	1	1	0	2	1
Total Recalls Conducted For Each Class and FM Contaminant		37	14	20	3	12	8	13	4
Percent of Total FM Recalls Represented by Each Recall Category and Contaminant			37.8	54.1	8.1	32.4	21.6	35.1	10.8

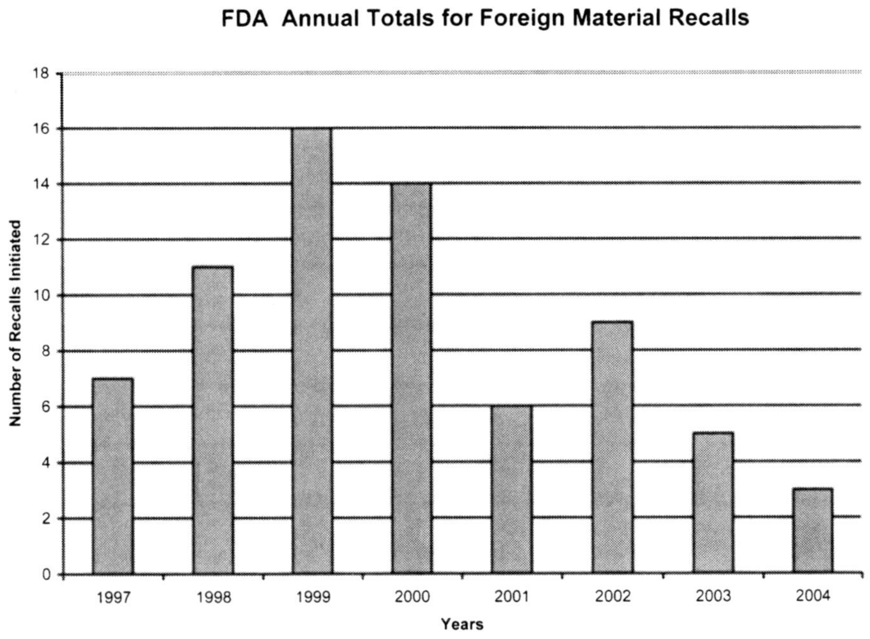

FDA Annual Totals for Foreign Material Recalls

Figure 1.6. FDA annual totals for foreign material recalls. Results from a 7.5 year retrospective analysis.

information that they may be privileged to and have kept confidential for good reason. It is interesting to note, however, the disparity in the number of Class I recalls for FM that each agency has classified. The only Class I recall from FDA was for a baby food product contaminated with pieces of hard plastic. Interestingly, FSIS conducted few Class III recalls for FM, and tended to cluster the actions into Classes I and II.

Obviously the increasing use of plastic in food manufacturing facilities and packaging components has been a common source of recalls for both agencies. As will be discussed in later chapters, the detection options for the presence of plastic in food products are quite limited, compared with those for metal and glass, so prevention strategies are the best approach for the processor. The recalls present in the FDA system, however, had a clustering (nine recalls) for metal wire being present in recalled products. Sometimes the wire source was from screening devices being used as a physical separation technique for other types of FM at the manufacturing facility or supplier location. The detectability of a metal wire component, discussed later in the text, is a common Achilles heel for processors so it was interesting to see it rank so highly in the number of recalls conducted.

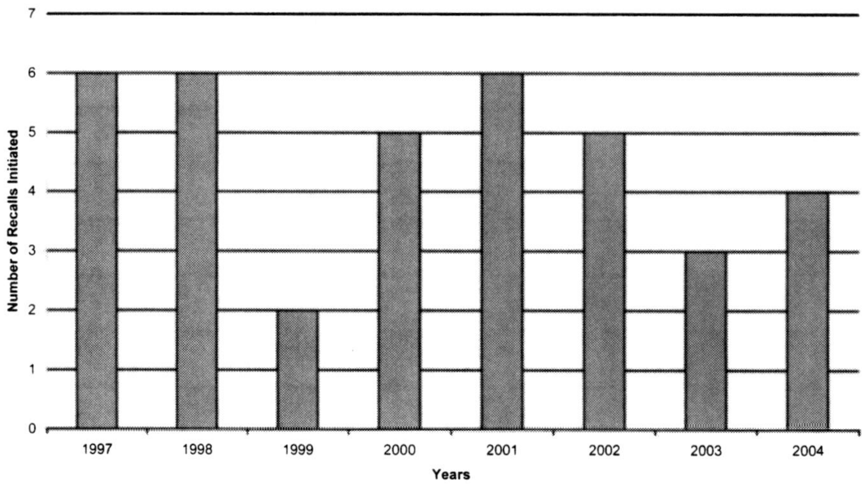

Figure 1.7. FSIS annual totals for foreign material recalls. Results from a 7.5 year retrospective analysis.

1.5 Why it is important to have a functioning foreign material control program

The data presented previously in this chapter clearly displays that FM is reasonably likely to be present within food products if precautions are not taken. There are essentially three general areas of reasoning that should compel processors to generate and follow FM control programs. The three "why" areas for members of companies to focus on controlling FM are:

1. Ethical
2. Legal
3. Monetary (Business)

The legal and monetary concerns can be argued as nonindependent by some; however, it is important to describe both areas in detail.

1.5.1 Ethical concerns

With the exception of terrorists, saboteurs, and others who are criminally insane, the practice of intentionally introducing FM into a food product or knowingly distributing food products that contain FM is contrary to human nature. Simply stated, the food processor has an ethical obligation to not knowingly distribute products that are unsafe, or otherwise misrepresent the contents of the product to the consumer. Those that argue against this statement, thankfully, represent an extremely slim minority of producers and their long-term existence is surely not going to be guaranteed.

1.5.2 Legal ramifications of nonconforming food products

The legal ramifications of distributing products that contain FM go beyond the sanctions that can be imposed against a company or individual by a regulatory agency. The United States Federal Court system, in concert with various state courts, provides an avenue for those harmed by a defective product to seek retribution. Once a plaintiff proves that their case has legal standing in a court of law (meeting the legal definition of the offense within the appropriate statute of limitations and there is sufficient evidence present to move forward), it is referred to as a tort.

The product liability laws of the United States have evolved substantially over the last century, and there is now a very clear path of recourse for a consumer whose health was deleteriously affected by use of a defective product. The number of product liability suits filed in Federal courts had increased nearly 1,200 percent between 1974 and 1990 (2). The grounds for pursuing action against a food manufacturer can be had through making a legal case for (1) Breach of Warranty, (2) Strict Liability, or (3) Negligence. The breach of warranty argument is the least utilized method for attempting to win damages for use of a food product. This is due to the fact that it would require the existence of a direct sale relationship between the manufacturer and the consumer (known as a condition of privity). Strict liability and negligence claims represent the majority of FM claims that would require a processor to defend itself.

The concept of strict liability was defined by the Supreme Court of California in its 1963 ruling of the case of Greenman v. Yuba Power Products Inc. Many other classic precedent-setting cases have occurred since the Greenman case, but all have shared the same basic components. The basic requirements for a strict liability case to be pursued against a manufacturer are: (a) the product must be new, tangible property (i.e., it wasn't something bought from a reclamation center or flea market); (b) the product must be defective when it leaves the manufacturer's control (this can include incidents that render the product potentially injurious within distribution centers); and (c) the product must be the proximate cause of the injury to the plaintiff.

A processor can also be held liable in tort law if proven negligent for injuring a consumer. Negligence cases rely upon the fact that reasonable care was not taken to prevent the defect from occurring in the product in the first place. In legal terms, as food industry professionals, our corporations are responsible for understanding the potential risks associated with the products and services being manufactured, distributed, and/or sold. The body of reasonable knowledge that a court of law will assume is possessed by a food processing professional is vast. Typical examples of reasonable knowledge can include the following areas:

1. Relevant article in mid-volume trade publications
2. Any published peer-reviewed scientific studies related to the hazard in a food product
3. Any documented recall cases (both past and present)
4. Practices used by others in the industry to mitigate hazards
5. Knowledge of all applicable regulations

When you or your organization make business decisions that are contrary to the contemporary body of assumed reasonable knowledge, the organization may now be exposed to excess risk in a product liability case. The punitive damage awards in a typical negligence case can be financially painful for an organization.

The disposition of product liability claims that are heard in a court of law has resulted in a 60 to 40 split for the defendant (2). It has been estimated, however, that nearly 95 percent of the lawsuits initiated by consumers against food manufacturers are settled outside of a court of law. A group of individuals from the Economic Research Service (ERS) of the USDA tracked various product liability lawsuits involving foodborne illness from 1988 to 1997 (3). They noted that in these gray area cases the plaintiffs were more likely to win if they could link the illness to a specific pathogen. Contrasting the concept of a foodborne pathogen that is difficult to isolate and culture to that of a FM which is highly visible and tangible to a layperson, it would be easy to see the difficulty in defending an FM case in a court of law. The median award for a plaintiff in the 175 foodborne-illness cases studied was $25,560. The mean award was much higher—$133,280—due to extremely large sums of money being awarded in a few cases (3 and 4).

1.5.3 Monetary concerns of the business venture

Delivering long-term sustainable growth of profit is the desire of all food companies. Long-term value positions of the company's products to consumers can only be attained through providing products that possess tangible characteristics which differentiate it from a competitor's product. Undoubtedly any food processing company that has enjoyed long-term success has developed a host of core competencies to deliver the results. Among the pinnacle core competencies that are essential for commanding superior market share and margin contribution, over the long term, are those related to the effective management of systems to deliver food safety and quality attributes.

High quality, safe, and convenient food products are viewed by consumers of the modern era as an entitlement, not an optional trait. Many studies conducted in various countries have found that consumers are willing to pay a premium for safe and wholesome products (5). The competitive nature of the industry does not allow manufacturers of products possessing marginal safety and quality to afford long-term premium pricing or top-tier market status. Intense competition in many areas of the industry has brought rapid and rampant consolidation of brands to further leverage economies of scale, expand core competencies, and strengthen distribution networks. Companies that do not embrace or exhibit the basic core competencies mentioned above will, in time, cease to exist in this type of competitive climate.

The cost of conducting a recall of a widely distributed food product today can be quite exorbitant. A recall (even Class III) can mean financial ruin for a corporation that is not financially prepared. Numerous mathematical formulas exist to screen potential investments, and all use a common theme of what percentage rate of return or future positive cash flow will be attained over a period of time related to the initial capital invested. Any capitalist or entrepreneur would be amiss if the investment that they made was one with a high likelihood of poor return.

As we examine how corporations and small companies attain funds for business activities, there are only two types of sources that need to be discussed: internal and external. Internal sources of funding originate from profits gained through business activities, whereas external sources of funding largely come from individual investors, corporate entities, or loans from financial institutions. When examining the ways to increase a company's profits, the most notable two center on:

1. Decreasing fixed and variable costs associated with the manufacture, sale, and distribution of a product

or

2. Increasing the sales quantity or sale price of a unit

Most corporations tithe percentages of their net profits into future-looking entities such as research and development, marketing activities, and general market surveillance. Such activities are important because they can help to increase the amount of future profit or thwart the competition from eroding the existing profit margin. Renewed investment by the groups mentioned above is what has allowed novel technologies to gain a foothold in the industry. Investment in novel processing technologies affords increasing levels of automation in manufacturing facilities, changes in packaging formats (such as the increased use of flexible packaging), and product quality improvements.

All of these components can help a corporation to increase their profitability or sustainable competitive advantage in a competitive market. It is difficult to successfully increase or sustain the profitability of a corporation if the monies and finite resources of the corporation meant for future-looking activities and novel technologies are being utilized elsewhere. The legal defense of FM incidents, the conductance of product recalls/stock recoveries, and the performance of costly reinspection activities are not value-added propositions for a company to undertake. It is best to build a framework of proactive strategies for controlling FM issues and rely upon a systematic approach to implement these. Proceeding forward in any other manner is akin to rapidly driving a vehicle down a busy interstate highway using only the rearview mirror for sight and directional cues. This can be quite challenging and energizing for the driver, but in time a catastrophic collision will happen.

Works Cited and Select Resources

1. Alli I. 2004. *Food Quality Assurance: Principles and Practices.* Boca Raton, Fla.: CRC Press. 176 p.

2. Brand M. Keller G. 1995. *Note on Product Liability.* Boston, Mass.: Harvard Business School Publishing Corporation. P 1–5.

3. Buzby JC, Frenzen PD, Rasco B. 2001. *Product Liability and Microbial Foodborne Illness.* In: Economic Research Service (ERS) Agricultural Economic Report No. 799. Washington, D.C.: U.S. Dept. of Agriculture. P 68.

4. Buzby JC. 2001. *Consumer Food Safety Behavior: Restaurants the Chief Target of Foodborne Illness Lawsuits.* In: Economic Research Service (ERS) Agricultural Economic Report No. 799. Washington, D.C.: U.S. Dept. of Agriculture. P 4.

5. Caswell JA. 1995. *Valuing Food Safety and Nutrition.* Boulder, Colo.: Westview Press. 457 p.

6. *FDA/ORA Compliance Policy Guide*—Chapter 5. Subchapter 555. Section 555.425. 1999. Washington, D.C.: U.S. Food and Drug Administration.

7. *Food-at-Home and Away-from-Home Inflation at 2.1% for the Year.* Jan 19, 2004. Elmwood Park, N.J.: The Food Institute Report (3):1–3.

8. *Food Price Outlook 2004 Updated.* Jan 1, 2004. Elmwood Park, N.J.: The Food Institute Report (1):1.

9. *Food Safety and Quality: Uniform, Risk-based Inspection System Needed to Ensure Safe Food Supply.* 1992. Washington, D.C.: U.S. General Accounting Office. P 8.

10. *Food Sales Clubs & Supercenters Eat into Supermarket Sales; Clubs & Supercenters Accounted for 5.8% of 2003 Retail Trade.* Feb 16, 2004. Elmwood Park, N.J.: The Food Institute Report (7):1.

11. Kramer A. 1973. *Food and the Consumer-Revised Edition.* Westport, Conn.: AVI Publishing Company, Inc. 256 p.

12. Olsen AR. 1998. *Regulatory Action Criteria for Filth and Other Extraneous Materials.* Regulatory Toxicology and Pharmacology (4):181–189.

13. Olsen AR. 2003. *Ouch! Regulatory Perspectives on Foreign Objects in Food.* Food Quality (3/4):63–64.

14. Olsen AR, Zimmerman ML. *Hard or Sharp Foreign Objects in Food.* 2002. In: Food Plant Sanitation. Pederson, N.Y.: Marcel Dekker Inc. 752 p.

15. *Presence of Foreign Material in Meat or Poultry Products.* 2003. FSIS Directive 7310.5. Washington, D.C.: U.S. Dept of Agriculture. P 1–7.

16. *Record-breaking New Product Activity in 2003.* Feb 2, 2004. Elmwood Park, N.J.: The Food Institute Report (5):1, 5.

17. *"Real" Sales at Supermarkets and Restaurants in 2003; Post Largest Gains Since 2000. Jan 26,* 2004. Elmwood Park, N.J.: The Food Institute Report (4):1.

18. Sinclair, U. *The Jungle.* 1906. New York: Doubleday, Page & Co. P 162.

CHAPTER 2

Proactive Management Strategies for Dealing with Foreign Materials

2.1 Introduction

This chapter is not meant to serve as a regurgitation of the seven principles of HACCP, or be filled with all the relevant GMPs to the food industry. Instead, this chapter is designed to be a discussion of the other systems and components that successful foreign material control programs are composed of. Obviously HACCP and GMP compliance are concepts that should be embraced by any food processor, but processors who only practice these systems may fall short in providing for general quality concerns. HACCP and GMP alone do not give instructive guidance on managing the risks present in ever-evolving and convoluting supply chains. The goal here is to describe, in detail, what the systems should look like in order to better control foreign material incidents within a supply chain. Consider this chapter a vision of what the high-level support systems must look like to deal effectively with potential FM from farm to table.

2.2 Step 1—Learn to recognize and emerge from firefighting and continual crisis management

The largest barriers usually cited by processors attempting to implement the practices presented below, and throughout this text, are the availability of resources and time. There is a simple reason for this that is more deeply rooted than ever-shrinking personnel or budget cuts. This reason is found in the fact that there are many times a lack of clear leadership priorities present within organizations. When all of the finite resources of an organization are dispatched for the practice of firefighting issues as they occur, there will never be any time dedicated to systems building or continuous improvement. Implementation of the practices presented in this text must be prioritized against day-to-day work activities that afford little evolution from the status quo. Crisis management (as presented in later chapters) is never something that should be performed half-heartedly, but to emerge from the crisis management mode an organization must focus its resources on developing systems for prevention.

Leadership commitment must be achieved and refocused on addressing FM incidence via use of a systematic approach, not via expending efforts and dollars to mop up after an incident has occurred. Many times personnel within the accounting and customer service functions will be able to assist in gathering this type of information. A majority of people in quality assurance and product safety functions typically fail to present the need for support of FM control programs to business leaders because they only cite relevant regulations and product specification requirements. To attract the attention of business leaders (who typically broker resources to relevant causes), it is important to present a need that is relevant to the bottom line. Typically a proactive manager would need to spend time tallying the monies wasted over the past 12 months in dealing with the presence of FM, or returning to steady state after a large incident. Typical areas to look for costs that can be included in a management review should include:

1. Detailed estimate of any monies lost in yield from the process due to sorting or culling activities that routinely take place as part of normal operations (kickouts, sortouts, scrapped batches, etc.)
2. Estimate of expenses for replacing equipment that is damaged or prone to excessive wear due to the presence of FM (knife changes, broken conveyors, broken grinders, over-amped motors and drives, etc.)

3. Additional clean-up costs incurred to deal with the FM [equipment disassembly, additional clean-in-place (CIP) or clean-out-of-place (COP), and increased chemical usage]
4. Total cost of product that was destroyed or donated for not complying with established FM specifications or policy. These costs should be calculated in the fully absorbed state including direct labor, indirect labor, materials, utilities used, storage costs, etc.
5. Amount of product (dollars and quantity) that is currently not shippable due to FM reasons, and the carrying costs associated with that product (quarterly tax value, storage, handling, etc.)
6. Opportunity cost for any missed sales during customer outages due to product not being shipped for FM reasons
7. Cost of any rejections or returns from customers or consumers and any reclamation fees charged
8. Cost of handling consumer complaint cases (correspondence, product replacement, medical expenses, and attorney fees)
9. Inspection or rework costs for dealing with product potentially not complying with FM standards
10. Cost of any field corrections, market withdraws, or recalls conducted for FM issues

The processor can commence assessment and system-development activities to prevent FM contamination of foods once an adequate business case is made for dispatching resources to the issue.

2.3 Step 2—Examine the existing system's infrastructure, and shore up shortcomings

Possessing specifications, policies, and procedures that address FM is an excellent first step to using a documented quality system as a defense. Simply stated, no food product should be manufactured on a large scale nor purchased in significant quantities without formal written specifications and documented quality systems. The relevant specification should include FM that is both product related and nonproduct related, with maximum thresholds clearly identified. The specification itself should represent the quantitative technical interpretation of the requirements needed to satisfy:

1. Customers
2. Consumers

3. Regulatory agencies that govern its manufacture, sale, or distribution
4. Any requirements for an internal downstream operation
5. Financial need

When thinking of the control of various FM. it is easy to see how items like metal or fibrous plant material could fall into multiple categories. Obviously certain types of FM will have a zero tolerance window (as was presented in Chapter 1), but some types may have a definable quantity per regulation, brand identity, or for the purpose of differentiation.

Development of clearly written specifications allows for broad communication of the product requirements. This communication can be for internal purposes, or for the purpose of engaging in supplier relations. Without a specification, it is very difficult to develop and maintain a stable business relationship with a third party (regulator or auditor), or communicate technical product characteristics to a potential customer. Regulatory agencies inquiring about FM practices, while not privileged to a review of your specifications, can be sometimes appeased by knowing that in fact they do exist.

2.4 Step 3—Apply source control strategies

Source control describes the condition where FM is contained prior to reaching the processing facility. Due to the nature of agricultural products and materials purchased on the commodities market, source control can be a difficult practice to achieve. Effective source controls, however, are by far the most effective means of ensuring control over FM. If the FM never reaches the processing facility, it would be very difficult to pass it on through the distribution channels. A processor can generally apply source controls in two ways.

The first source control scenario can be employed by processors who purchase raw agricultural commodities or livestock that are to be further processed internally as a value-added measure. In this situation, the source control strategy for FM will focus on many aspects of current good agricultural practices (GAPs) that must be practiced by those involved from the farm through receipt at the processing facility. The level of GAPs required to prevent FM are mostly based on common-sense principles and applied knowledge of preventing potential FM that is characteristic of the commodity. The most common GAPs that can be scrutinized are presented in the following section.

The second form of source control revolves around the need to manage suppliers and brokers who vend raw materials to the facility. There are many techniques and concepts that can be used by the processor to make supplier management a very effective form of source control. When introduced in a systematic manner this type of management strategy can be implemented with few additional resources, but it requires discipline. Supplier management source control strategies for FM are presented later in the chapter.

2.4.1 Source control 1A—Application of relevant GAPs for raw agricultural commodities

Ingredients of plant origin require a different strategy than those of animal origin due to the nature in which they are grown, cultivated, stored, and transported. Livestock and chemically synthesized ingredients present different contamination potentials from these raw agricultural commodities because their production and sourcing are inherently different. To better understand what GAPs are of greatest importance, one must have a thorough grasp of the life cycle of these commodities. The sections that follow will illustrate the typical life cycle of an agricultural product, and highlight where appropriate FM controls are typically missed or not properly applied.

2.4.1.1 Geography and details about the crop production region

Obviously not every crop will be able to be sourced from a desirable region, but there are many screening questions that can be asked when dealing in identifying desirable raw agricultural products. The questions must identify the physical location of the fields that are harvested to yield the purchased product. A good start for understanding the FM and GAP controls that should be applied is to know the geographical region that the product will be grown within. For instance, if you were to purchase sweet potatoes grown in a region that is typically known to produce peanuts, there is already an FM control to ask about. Likewise, if the field areas were redeveloped from land once used for landfill or other industry, another potential contamination route can be identified. Field crops grown in areas that border busy public roads, railroad beds, or other traffic patterns for the general public are usually not the most desirable from which to source product. Think about contaminants such as beer bottles, candy wrappers, paper, aluminum, rubber, and various other metallic components in this scenario. At a minimum you should be able to identify what other produce is grown in the region, and what the area is like that borders the fields.

The facility layout is important as well. Does the grower employ personnel who are intimately involved in the field region? If so, are bathroom and eating areas designated? To control access, is there being kept a central check-in point or roster identifying who is working? Are the growing fields fenced in (unlikely) or wide open? Are mobile irrigation techniques used, or is fixed sprinkling the main method of providing additional irrigation? (Moving irrigation systems sometimes have a tendency to lose parts.) How many times would a tractor or pickup truck typically cross the field for the purposes of irrigation, fertilization, spotting, or application of pesticides or herbicides?

2.4.1.2 How the product is harvested and initially sorted

The nature by which the product is harvested can help the producer understand what types of GAPs may be necessary to deal with the FM that may be introduced by such practices. One universal constant the processor can rely upon is that field crops do not harvest and clean themselves when the time is right. The products also do not sort themselves into desirable or undesirable categories for use...this task remains for the grower or processor to complete. In general, the harvesting techniques used to procure agricultural products can be broken into two categories:

1. Manual harvesting
2. Mechanical harvesting

Obviously manual harvesting involving disciplined and guided human judgment will provide the highest level of discrimination in screening FM from a harvested product. Mechanical harvesting can be excellent as well, but its efficacy is dependent on multiple control factors and crop characteristics to remain reliable throughout the harvest. Currently the vast majority of agricultural products utilized in the food industry are procured via mechanical harvesting. The insurgence of organically grown and specialty boutique products has helped to increase the amount of manual harvesting, but this method still represents minority harvesting techniques by a wide margin. Regardless of the technique used to harvest the raw produce, a written specification with clear tolerances for the typical FM that will repeatedly be encountered in the product is a necessity.

2.4.1.2.1 Practices of concern for manually harvested agricultural products

Typical FM concerns associated with manually harvested crops deal with personnel factors. The level of training given to the personnel performing the harvesting activities can be a significant factor. Whether story boards are used to represent what acceptable grades of product are, or whether there is a hands-on briefing conducted in the native language of the personnel doing the labor, are two solid questions that need to be asked. Additionally another source of FM concern is the way in which personnel comply with such control strategies as (1) beard net/hair net use, (2) wearing clothing that is clean and not readily friable, (3) wearing jewelry or bringing personal effects along with them, (4) eating in designated areas, (5) use of designated toilet facilities, and (6) how are sanitary napkins and tampons addressed.

The method of payment used for personnel harvesting the product needs to be examined. For increasing productivity, a majority of growers pay their labor by a harvest quantity usually expressed in volume (bushel) or weight (kilograms). In both scenarios the practice must be examined to ensure there are controls for dockage of defectives. If there are not controls in place, then there will be a temptation to cheat the system. I personally witnessed an individual place stones in the bottom of a harvest bucket of strawberries to increase reward distributions in a pay-by-the-kilo operation. If the reward system is based on volume, expect that bulking agents like grass or other plant materials may be found as well.

The containers used for collection and intermediate storage of the harvested produce is an additional issue that needs to be addressed. Because of their weight and fragile nature, containers made from glass are almost never used for produce collection. Materials used on a more routine basis include cloth woven bags, nylon woven bags, plastic pails, metal pails, wheelbarrows (plastic or metal), and light-gauge plastic sacks. Each one of these collection containers presents a unique potential for generating FM. Knowing which type was used can help the processor hone in on identifying contamination, and apply separation methods. Obviously if the produce is going to be later washed and/or graded following harvesting, the chances of these types of FM making it to the processing facility are slim. Remember, however, that we are discussing prevention strategies that are proactive in nature, and that by applying relevant GAPs, source control can prevent the FM from becoming an issue in the first place.

2.4.1.2.2 Practices of concern for mechanically harvested agricultural products

The word "mechanical" instantly implies that machinery is used to some extent in harvesting the product. Any time there is moving equipment in proximity to a raw product, there is potential the equipment could seed the product with FM. Typical FM from harvesters include nuts, bolts, gear teeth, sheet metal, welding flash, broken tines, shattered plastic, lubricants, wire ties, wires, and glass. The age and condition of the equipment utilized for harvesting the produce must be carefully scrutinized. If equipment that appears to be falling apart while standing still is presented for use, an alternate grower/harvester may be a better option. Thankfully the likelihood of the presence of most of these contaminants can be addressed via use of physical separation techniques, magnetic separators, and metal detectors applied at various steps anteceding the harvest.

The other concern with mechanical harvesting is directly related to the lack of artificial intelligence of the equipment itself. Thankfully, in recent years, there has been a proliferation of metal detection equipment being incorporated into the design of harvesting machinery. The root of this design change was driven from an equipment protection standpoint, but should continue to offer tangible benefits for preclusion of FM during harvest. Due to the indiscriminant nature of mechanical harvesting techniques, other undesirable FM may be taken from the field in addition to the produce itself. Stones, bones, dirt, weeds, stems, stalks, roots, leaves, snakes, reptiles, rodents, and other various FM can be taken up by the harvester if not properly operated. Use of spotters and air selection at the point of harvest is critical for the harvest of many common commodities versus just taking up the field and sending it on to the grading facility or processor.

2.4.1.3 Graded and nongraded produce

The grading of most raw commodities is performed using physical selection techniques such as weight, color, and size of the product presented. The products are discriminated into unique classifications (grades) based on these inherent characteristics. Other specific factors for each produce item such as the extent of rot or the presence of extraneous matter are usually included in the grading process as well. For this discussion, the notion of purchasing graded produce versus nongraded produce items will be discussed. Tips on ensuring that the grader is operating properly, if the produce was graded, are also presented.

Purchasing graded produce or sort-outs from a grading station usually can provide another level of reduction in FM. This assumption must be affirmed by confirmation that the grading station is running in accordance with standard operating procedures and the equipment is maintained properly. Usually the purchase of graded produce items comes at a premium for the producer versus buying the field in an "as is" state at harvest time.

There are several techniques that can be used to ensure that grading stations are operating properly. If the grading is done by size, for instance, simple analytical techniques can be used to ensure that you are not being sent products that may contain a size-sensitive FM. "Size sensitive" describes a situation where the undesirable FM of concern is mostly contained in the smaller fraction of product that isn't supposed to be in the grade that is purchased. If the oversized portion is purchased, a statistical sampling of loads of purchased product can be performed (see Chapter 9 for guidance), and the results plotted on a frequency distribution (histogram). Knowing that variation is always present in agricultural commodities can guide the interpretation of the results. There should be a normal distribution of product sizes present in the produce, and a sharp cut present at the declared undersize point. If there are large gaps between sizes, you can be assured that sorting of some kind has occurred. In addition, if produce is found outside of the established cut point, the FM of concern may be present as well due to improper operation of the grading station or wear/failure of the sizing grid. If this phenomena is recurring, an alternative grading station should be sought.

2.4.1.4 Storage, staging, transportation, and transfer of agricultural products

Another major concern for the introduction of FM into a raw agricultural product is found when transportation and loading practices are examined. Typically any product will be staged in containers of various construction following the harvest, cleaning, and grading process. Typically the material used in the construction of the container itself will be a variant of metal, wood, or hard plastic. The containers are meant to protect the food items from the environment, but can, and often do, become a source of FM.

It is important to realize that each type of container will present its own potential for introducing FM. Metal containers of appropriate construction represent a large financial investment for the processor. Metal containers are usually purchased for the purpose that they should stand the test of time. When corners are cut in construction or repair of the containers, however, FM can be the result. Unfortunately this can happen when incompatible grades of metal are used (such as galvanized steel), for construction or repair of the container (displayed in Figure 2.1). When this occurs the container can become an FM generator. Typically the container will have rust, peeling paint (paint is usually applied to hide the rust), or welding slag present from repairs that may become intermingled with the product.

Incompatible wood containers can present FM issues as well. Wood by nature may harbor various living entities across the evolutionary continuum, from bacteria to rodents. The wood itself can fragment into splinters or chunks that can become entrained in the product. The wood containers are typically held together with fasteners which can introduce metallic contaminants (such as nails and staples) into the product. Figure 2.2 displays the typical use of wood containers for harvesting apple crops.

Plastic containers can be a great alternative to wood and metal containers, but they too can present problems if incompatible handling practices or materials are presented. Plastic containers will create shavings or particles if improperly handled or manufactured from the wrong components. Rigid plastic containers also quickly exhibit fatigue, becoming brittle when they are used beyond their rated capacity or are subjected to extremes in climate (freeze-thaw conditions). Figure 2.3 displays a typi-

Figure 2.1. Nonappropriate metal transport containers for produce items.

Figure 2.2. Typical wood containers used in the harvesting and storage of produce.

Figure 2.3. Example of a rigid plastic container used for the harvesting and storage of produce.

cal plastic container used for the harvest of produce. Flexible plastic packaging presents issues with tearing and puncturing. Such plastic packaging can be found in the lining of ingredient drums, combinations of frozen produce or meats, corrugated containers housing dry or frozen ingredients, and super sacks. If the plastic lining tears via puncture or unloading practices, it is very difficult to identify it in the product (unless a leakage situation is identified during transport or unloading). Many processors, for this reason, specify the use of an unusual color for plastic linings to aid identification in the case of accidental contamination.

2.4.1.4.1 Staging of the product prior to loading and distribution

Not many farm fields have sufficient area under roof for handling the intermediate storage of products prior to initial shipment. Usually raw agricultural products are staged closely to the area that they were harvested in for loading and distribution. Staging practices and the areas used should be scrutinized to prevent introduction of FM. Figure 2.4 illustrates the typical staging practices used for sugar beets. Thankfully, the process of manufacturing sugar is quite robust in terms of culling foreign materials since the staging of the raw product presents great opportunity for contamination.

Figure 2.4. Storage conditions of sugar beets prior to being staged at the processing facility.

Imagine bins of harvested strawberries parked under a large shade tree. This scenario might seem desirable because it presents convenient shade for the produce, better ensuring that it does not get too hot and lead to the formation of rot. Just imagine the result, however, if the shade tree was a walnut variety (allergen), and/or if it was filled with birds or other furry mammals—a naturally good source of hair, feathers, beaks, and bones FM when they enter food processing equipment.

Other considerations for the staging of agricultural products are access controls and length of time exposed. The staging area for the product needs to be examined for unauthorized pedestrian access or other undesirable traffic patterns (food security threats). The length of time the material is left in a staging area is important as well. The probability that an inadvertent FM may be introduced increases with the time of exposure . . . this is especially true if the product remains uncovered during this period.

2.4.1.4.2 "One if by land, two if by sea" loading and transportation of the product

Be wary of products that have traveled a long way to reach the facility. The road less traveled is the preferred product distribution path for preventing incidental contamination of FM. Over-the-road transport of products should be done with the product secured in a manner that FM cannot be introduced inadvertently. Ensuring that tank lids and pumping outlets are secured, tightened, and sealed on tanker trucks is an important practice for bulk transport of liquid or solids. Any external piping connections should be capped and affixed with a seal as well. If the materials are housed in bulk bins with open tops (apples, for example), the load must be tarped. If the material bins are meant to be stacked for transport, the bottom of each bin needs to be cleaned of FM prior to the commencement of stacking. This is true of any surface that could retain FM that may fall into the bin adjacent or below it. Interlocking the bins to unitize the load is critical as well. During a long highway excursion, the bins of product may rub against one another in a manner that creates shavings or flaking materials that could become commingled with the product. By using bins constructed of appropriate materials and utilizing proper unitization practices for loading trucks, this scenario can be avoided for the most part.

Railcars and shipping containers should be treated similarly to over-the-road transportation, with some minor exceptions. Dunnage will typically be used and reused in situations where the products may shift during transport. The dunnage could be constructed of corrugated wafer-type materials, or simply be plastic bags filled with air. These materials may come into contact with the containers or product itself, so care must be taken during loading and unloading of the car or container. All cars and containers should have serialized seals attached to the secured doors, hatch openings, and/or valves for piping. The other issue with transport via rail is that the locations where the product may sit for extended periods are usually unknown by the shipper or receiver. Cars and shipping containers may sit in switchyards or at other locations for weeks while being unloaded, transloaded, or transferred to another spur. Proper container designs that do not allow unabated access to the product are a must for situations such as these. Solid walls, functioning door seals, and solid hinge design are all critical aspects to limit access by unauthorized individuals.

Regardless of what type of transport mechanism is used, it is critical that the area housing the product is verified to be free of filth and other FM impurities before it is loaded. It is a good policy to require the transportation company to provide traceability of what was last hauled by the transport vessel being loaded. Obviously a live bottom truck that carried wood waste from a paper mill or compost from the local sewage treatment plant on its last trip shouldn't be considered for transport of sugar beets or carrots. Likewise, wash certificates should be requested for bulk tankers (truck or rail) used for transporting liquids. Fresh-pressed grape juice can pick up an odd hue when it is put into an unwashed tanker that had hauled yogurt on its previous trip. Bulk dry pneumatic trucks used for the transport of media such as grains or salts need to be carefully examined to ensure that the intake air is filtered, and the filter is of appropriate construction to keep FM from entering the product line.

When products are unloaded at the facility, care must be taken to examine the containers for any signs of visible damage or evidence of tampering. Before hooking up hoses to tankers (pneumatic or liquid) for transfer, the person responsible for off-loading should use a flashlight to inspect the throat of the piping (which should have been capped and sealed before transport) for the presence of any potential nonrelated FM. Piping used to transport sugars and starches is especially attractive to nesting insects and rodents that might not mind going over the road. The hook-up connection area should also be inspected for galling or flaking metal, friable rubber gasket material, or plastic seals that are cracked or indicative of excess wear. If the product being received was sampled at customs, ensure that the paperwork is official and in good order. Check to ensure that the dates match the travel period, and that the new seals applied to the load or container match those referenced in the customs paperwork.

2.4.2 Source control 1B—application of relevant GAPs to livestock commodities

The major livestock commodities can be grouped into the categories of cattle, swine, and poultry. When actively participating in the production or acquisition of livestock, there are several factors that need to be identified to ensure that FM is contained. Much like raw agricultural products, livestock is a commodity that doesn't just by chance become a sliced or processed ingredient of optimal quality. The birthing, herding, feeding, and ongoing health promotion of the livestock are critical for ensuring that they will be fit for slaughter and subsequent processing. There are periods of time along this continuum where FM could be introduced, and these situations can be managed by applying the appropriate controls and monitoring practices.

2.4.2.1 Regional considerations

Where is the livestock raised and housed? This question can offer important insight into the types of environmental influences that could result in FM issues later. There are still certain regions of the world where the technique used to herd or round up the animals is to shoot at them. This is typically done with a pistol which contains cartridges loaded with small lead or steel pellets. Depending on the proximity of the animals to the herder and the type of round used, the pellets can puncture the hide and become imbedded in the meat. The USDA and the industry readily recognize this type of contamination, and is quite weary of meat or livestock procured from certain regions of the United States and select foreign entities. Likewise, tranquilizer darts have been found imbedded in animals at the point of processing. Obviously the use of a tranquilizer gun is sometimes essential for disgruntled animals, but the dart should always be retrieved regardless of regional practices.

Certain areas of the world are also likely to provide pathways for infection of the livestock with certain parasites. This can be due to the quality of the water supply or other potential vectors being present in certain areas. From a HACCP standpoint, however, this type of issue is usually considered as, and handled as, a biological hazard by most processors. The concept of a worm or multiple 10 mm hydatid cysts (such as those produced by *Echinococcus granulosus)* being imbedded in a piece of steak, liver, or ham is not attractive (6). If the consumer can recognize the presence of such a burrowing worm in the product, they will not care that it no longer is of an infectious state...they just know that it is revolting to them.

2.4.2.2 Dietary considerations and captivity feeding versus free range

Grazing ruminants, such as cattle, goats, and lamb, are amazing creatures that can acquire nutrients readily when allowed to fend for themselves in pasture. Cattle are especially nondiscriminating eaters. Cattle do not usually prehend their food with their lips, and their smooth tongue affords FM a quick path into the digestive tract. When this occurs it is referenced as "hardware disease" because many times the contaminants become lodged in the rumen. Given time, sharp contaminants like hard plastics, nails, screws, staples, and barbed wire can penetrate through the abdominal cavity. When this happens the material will become imbedded in another part of the body, and may even find its way to the heart itself (5).

To combat hardware of the ferrous variety, most farmers who tend dairy cows and cattle will "give the cow a magnet" at an early age. The magnet administered is an Alnico permanent magnet pill (bar magnet about 2.5 inches long and ½ inch in diameter). The magnet, displayed in Figure 2.5, is administered to the cow much like any other pill or medication via use of a balling gun or pill forceps. Once dropped behind the tongue, the cow swallows the magnet, which will remain in the rumen for the rest of the cow's life and function to catch most of the ferrous FM ingested by the cow. By holding the material at its surface, the cow is protected from having sharp metallic objects migrate beyond the rumen. Obviously it is important for the processor to carefully remove the cow magnets following the slaughtering process. Improper handling of the magnet and rumen could lead to potential contamination of the surrounding meat.

As a preemptive measure, the feed for cattle should be carefully examined for purity. Feed bunks should incorporate magnetic separation and physical separation equipment (1). Even the crudest of screenings will help ensure that the chance of feed containing contaminants is unlikely. This is very important because the cow magnet will not catch some stain-

Figure 2.5. Example of a 6 cm Alnico "Cow Magnet" used to treat hardware disease in ruminant animals.

less steels, aluminum, plastic, or glass contaminants that could be present in the feed source. When cattle are out to pasture, there is a chance that they could still consume FM. Knowledge of the use history of the field is again important for determining the risk. If there is concern, however, the main roaming areas may be occasionally combed with a large permanent magnet or swept with metal detectors to ensure that FM is not present at the surface.

Poultry, especially chickens, can suffer the effects of a maladjusted feed mix. Egg-laying hens and breeders can be susceptible to FM contamination through the dietary mode. Usually egg-laying hens are slaughtered when their egg production slows or when feed prices escalate. In times like these producers may skimp on the feeding intervals for these animals (a poor practice), and the hens react by turning over their bedding material in search of nutrition. Many times the bedding material or other FM is swallowed by the chickens; the wood or straw can penetrate the esophagus of the chicken and contaminate the neck meat. This fibrous contamination should not be present in hens that are fed up to the point of transfer to the slaughter operation.

2.4.2.3 Caution when administering injections to large animals

It is often necessary to administer vaccinations or other medications to animals such as pigs and cows. The predominant means of doing so is through use of various-gauge hypodermic needles. If the animals are improperly restrained or a bent needle is used for the injection of medication, there is a high probability that the needle may break off underneath the hide. The injection site needs to be carefully considered as well, because injection over a bony area can also lead to a broken needle. Once the needle has detached and is inside the muscle, the movement of the animal will cause the needle to migrate. This never-ending migration of the metal needle makes extraction difficult for producers and veterinarians alike (3).

The problem of broken needles is quite common and the National Pork Producers Council launched a national campaign in 2001 to increase awareness of the issue. The campaign was entitled "One is too many" and focused on reinforcing the importance of producers having standard procedures for injection scenarios (11). The Canadian Cattlemen's Association has also helped to shed light on the problem of needle contamination in beef products. The association surveyed processors and veterinarians to gain a better understanding of the scope of the problem. Interestingly 41 percent of the processors surveyed (including packers, purveyors, and retailers) had received between 1 to 12 needle complaints within the last year. When the veterinarians were surveyed, 25 percent of them responded that they had experienced a problem with needle breakage at some point in their careers (11).

Most grades of needles used at the farm level would not lend themselves well to metal detection. Recently needles have been developed claiming that they are 100 percent metal detectable at high speeds, but no confirmation data was readily available at the time this text was assembled. Based on current knowledge of metal detection technology (as discussed in Chapter 5), the orientation of traditional needles as they pass through a detector will be critical in assuring they are rejected. The results of a study performed at Iowa State University (ISU) comparing the strength and detectability of various needles manufactured for veterinary use was released in 2002. The study was overseen by Steven Hoff, an associate professor in ISU's Department of Agricultural and Biosystems Engineering, and proved there was high variation among commercially available needles when the data was ranked for each style by manufacturer. The most interesting portion of the test dealt with the detectability of needle fragments in a 3-lb pork roast. The position effect was quite significant, and needle fragments passed through the detector at a horizontal position were more readily detectable. Overall the detectability of the test needles averaged out at 14.9 percent for the group (with some needles having much higher rates of detection).

Adopting proactive management strategies (including needle-use policies) and injection procedures will go a long way toward ensuring that the consumer will not be at the mercy of the metal detection characteristics of the needle itself. The use of alternate dosing such as oral administration or anal suppository should be pursued to avoid risk of breaking a needle off in the animal. Other good practices include positively identifying any animal that has had a needle imbedded within its muscle that has not been removed. Such traceability procedures will provide the processor with the knowledge necessary to take precautions during the slaughter process, or opt not to use the animal for human consumption. Producers should have a training program to accompany their injection and needle-handling standard operating procedures. By training the personnel who will be hands-on in the area, the chance for stray needles being left out, or unreported breakages, should significantly decrease.

2.4.3 Source control 2—supplier management practices

Obviously not all components used to manufacture food products will be directly sourced from the agricultural community. Complex ingredients such as spices, oils, flavorings, vitamin complexes, and preservatives are going to be sourced through external producers specializing in each component. Other components used in food production such as food packaging materials, sanitation chemicals, and lubricants will also be purchased from specialty manufacturers or brokers.

The contemporary quality assurance philosophies presented by Deming and Juran helped to abolish the draconian inspection-based quality control (QC) cultures that existed across industrial society. This enlightening period yielded a decrease in emphasis on inspection of purchased commodities in lieu of developing auditing systems and supplier certification programs. Testing was still required to some extent for identifying gross excursions or trends, but it was greatly reduced from elevated inspection levels over time. The quality assurance approach is proven, and works quite well. In today's environment of ever-decreasing margins on products, there are great pressures to reduce manufacturing conversion costs. Many of the containment measures employed by companies have forced audit frequency inspections to be reduced to extreme minimums. Now, more than ever, producers are at the mercy of their suppliers to assure the quality of goods purchased. This is especially true in the case of FM concerns. As noted earlier, most FM contamination incidents are present at either a very low level across a large amount of product, or at high levels with a relatively low frequency of occurrence. Routine audit sampling will not be sufficient to identify these foreign material incidents, so the controls in place at the supplier location are definitely key for the ongoing success of the brand.

As stated previously, a written specification must be in place for any supplied component, and it should include tolerances or other references to FM. Management of supplier contracts as well as the overall dynamic nature of information sharing are critical to the success of a foreign material control program. The sections that follow will outline the appropriate proactive measures to take within the life cycle of a supplier-purchaser relationship. Obviously, some of the concepts presented below will not be immediately actionable for processors that are already "locked in" to long-term supply agreements. The monitoring and feedback section, however, is a universally applicable step and should be initiated immediately.

2.4.3.1 Supplier selection

Organizations that select suppliers using the single component of lowest cost are hopelessly inept. The total value analysis concept has been validated through repeated study of the business practices of thousands of major corporations that are successful in the long term and exhibit continued growth. Organizations that do not embrace this concept will find that the few tenths of a cent saved on a per unit basis will quickly be lost upon the first lot rejection, customer service outage, or recall that results from such bottom feeding practices.

Much homework is required when selecting a new potential supplier, and this is especially true when the processor is concerned about FM controls. When evaluating new suppliers a simple screening questionnaire can be sent out with the bid package. The bid package must make completion of the questionnaire a requirement for review of the bid. Contact information for a technical resource that can clarify any of the questions found on the survey should be printed on the first page as well as clear instructions for submission. The questionnaire should ask the potential supplier to detail out certain elements that indicate they have a functioning foreign material control program. Obviously the scope of the questionnaire could include other points relevant to food safety or quality assurance, but for the narrower topic of FM control these will be omitted in this text.

The questionnaire should ask specifically for information on the processes used to manufacture the product, the ingredients used, and the sourcing strategy for these (internal, broker, or supplier). Typically a process flow diagram should be requested as well as a copy of the supplier's HACCP plan. Current sampling procedures, frequencies, and hold procedures should also be requested from the supplier. Another essential element of the questionnaire is to ask if their products are currently implicated or have been identified as the cause of any market withdrawals, stock corrections, or recalls. A list of other customers that the supplier services should be requested, and contact information for references that can attest for service and quality. Obviously many suppliers will object to providing all of the information asked in such a questionnaire, or cite proprietary reasons for leaving sections blank. Remember that the suppliers who at least make an attempt to fill out the questionnaire or engage in conversation about the topics presented are usually far better than those who do not respond at all.

Once the paperwork phase has been completed and the field of potential suppliers has narrowed down to a manageable number, many processors will want to perform a site visit or audit. For low-risk products the necessity of this step can be debated; however, there are alternatives to expending finite resources on a walkthrough of a process. Using a qualified third party auditor to evaluate a facility is usually a good option for a producer who is short on personnel. If the supplier location has been audited by a food-safe auditor from the Food Products Association (FPA), this is even better. [FPA is the newly named former National Food Processors Association (NFPA)]. Once the supplier gives permission to the purchaser for review of the audit results, a small fee is charged and FPA releases the results to the processor for review. The credentials and training required of an FPA food-safe certified auditor are quite high, and the audits are usually well detailed. The audit should be a good representation not only of the FM controls in place, but also the entire food safety system functionality.

2.4.3.2 Setting clear expectations

The tone of the supplier relationship should be set at the selection stages, and expectations on foreign material control must be clearly communicated and documented. Remember, if the supplied product contains significant FM, that just makes the processing activities needed to take place in your operations that much more difficult to achieve.

When possible, push back a great deal of the foreign material control strategy to the supplier base, but be realistic about expectations. Expectations pushed back on suppliers should not exceed what current generally accepted process technologies can reliably deliver. An example of an unrealistic expectation is one that specifies zero microscopic metal fragments allowed in any sample taken from a heavily processed ingredient. Some regulatory officials and purists may wince upon reading this statement, but microscopic trace purity analyses of practically any processed product will reveal insignificant quantities of microscopic metals, rubber, and other extraneous matter. Obviously the specifications issued to your processors should never approach or exceed regulatory allowances (DALs, or standards of identity). Such a policy would not only be considered illegal, but could result in significant risk to the established brand equity.

2.4.3.3 Alignment of test methods

There is an additional step that must be performed when settling on specifications for FM with your suppliers. This step is often forgotten by most processors and can result in significant decay of the newly forged relationship with their supplier. When finalizing the specifications, it is important to agree with the supplier in writing what method of testing will be used for identifying FM. The method referenced should, in most cases, be a published method that is widely used throughout the industry. Usually methods published by AOAC International (formerly the Association of Official Analytical Chemists), ASTM (formerly the American Society for Testing and Materials), and regulatory agencies are a good choice due to the qualification processes that these groups require.

An example of where failing to agree upon a method can be problematic is in testing for the percentage of bone present in a purchased ground meat product. There are multiple ways to determine the quantity of bone present in ground meat products. With the lack of a common method, the supplier may adopt a generally accepted method like the commonly accepted USDA test for the presence of bone. Simultaneously, the producer may adopt a test method that is of higher discrimination (like analysis of trace minerals such as calcium via use of an expensive atomic absorption spectrophotometer) for more precise analysis. Obviously both methods will produce data, but which method was in mind when the specification was agreed upon? From the author's standpoint, and in a court of law, if a consensus for the method to be used was never documented in the specification it could be a point of contention until the next supply agreement is negotiated.

An additional point that needs to be clarified is that the method referenced in the specification should also illustrate sampling techniques and statistical methods used to determine the number of units sampled. There are many certificates of analysis issued to food processors on a daily basis that are derived from an N = 1 basis (only one sample tested). This type of sampling plan information is essential when setting up purchasing agreements for packaging materials. Usually the processes for manufacturing packaging materials are quite elaborate, and the procedures used for sampling need to ensure that the sampling is statistically stratified. The sampling procedures agreed upon should include equal probability of testing across molds and mold cavities used for production. It is quite disheartening to chase an FM issue to root cause stemming from poor packaging supplier practices. Sometimes one can even find that the sampling that was "agreed upon" was understood to be on a line-by-line basis at the supplier's location. When this happens certain portions of lots may not have even been inspected, but rather combined with a known "good" lot and shipped.

2.4.3.4 Initial monitoring and routine monitoring of supplied product

When entering into a new relationship with a supplier, some extra effort spent on initially verifying their capability should be made. Hopefully, at this point, certificates of analysis (COAs) and/or certificates of conformance (COCs) are received by the processor for each lot supplied. An ongoing letter of guarantee sent via the supplier identifying the specific product that it relates to should also be on file by this point at the processing facility. Random statistical lot inspections for FM and oversight of production periods using the newly supplied lots are prudent first steps. Taking steps such as examination of the rate of process rejects (tailings, metal detector kickouts, x-ray rejects, magnet finds, etc.) are usually areas on which to focus resources during the early stages. The finds from these studies should be enumerated and mathematically reviewed versus the specification and the COA to determine if a deviation is present. By being proactive and diligent, FM issues can usually be contained at the early stages before they reach an extremely costly nature. The statistical sampling chapter (Chapter 9) presents several examples of supplier monitoring practices that can be readily applied by the processor so that meaningful conversations can be had with the supplier.

Conversations should be initiated with the supplier at this point to reaffirm the agreed-upon expectations and understand if the supplier is running into issues achieving this status. Initiating such conversation could make the processor privy to issues that otherwise may go unaddressed in the short term, potentially leading to a supply outage for suspected FM. Presenting documented feedback to the supplier (for both positive and negative events) needs to become an ingrained process to strengthen the relationship at this early stage.

2.4.3.5 Dealing with nonconforming shipments

An overriding internal policy must be in place for the processor to deal with nonconforming shipments. This policy should be approved at the highest levels of the company's functional management to ensure that all business areas (purchasing, operations, sales, quality assurance, product safety, etc.) are in alignment. The policy issued should clearly reference relevant rejection procedures and hold procedures, and define the authority levels required for granting the use of nonconforming lots. Education of the policy components and its intent is extremely important because, often in the absence of a well-communicated policy, the company will assume the risk for the supplier by using the materials. Typically this will happen when personnel in operations will attempt to cull the FM from the poor quality lots in an attempt to support the perceived business needs, achieve committed manufacturing volumes, and subsequently satisfy the needs of customer service.

The policy presented must also clearly state that if any deviation from the established specification is discovered, regardless of severity, it must result in a documented nonconformance being issued to the supplier from the processing facility. This document should contain all the relevant information necessary for the supplier to understand the scope and severity of the situation. Good examples of what types of information should be included on the nonconformance document are:

1. Name and contact information for the originator
2. Description of the material supplied that is suspect (supplier description and relevant code)
3. Any lot information that is relevant to the supplied material (date of manufacture, lot numbers, pallet tags, etc.)
4. Brief description of the specification exceedance and nature of the FM (contaminated with grease, shavings present, suspended mold mat, etc.), and whether samples are en route to the supplier. A picture or photograph of an example would be ideal as well.
5. The quantity of the lot already used in production
6. The quantity of the suspect lot still in storage

7. How much of the storage quantity is on hold
8. How much finished product is on hold (quantity and approximate dollar value) because of the nonconformance
9. Whether or not this hold will result in a material shortage in the foreseeable future
10. Priority of this issue for the business (critical, major, or minor, for example)
11. Expected date when a written corrective action is to be received by the processor from the supplier

It is critical that the supplier be held accountable for responding to the nonconformance with an appropriate corrective action to the situation. The corrective action provided as a response should be auditable, and deal with correcting the root cause of the problem. Responses such as "replaced suspect product with good product and fixed leaking hydraulic hose over transport conveyor" do not provide assurance that the irreversible corrective action for the root cause was made. Review of nonconformances issued to a supplier and the corresponding responses received over time can be a great tool when trying to identify suppliers with poor or deteriorating FM controls. These reports can paint a picture of whether the process improvements have been systemized or if the situation was dealt with in a one-time fashion. Nonconformance documentation also proves invaluable when it comes time to negotiate the next supply agreement or nominate a consolidation strategy for a supplied material. The price of a supplied item is always open for negotiation in such circumstances, but the supplier's past performance quantified with solid data and facts usually is not.

2.4.3.6 Rewarding exemplary suppliers

It is warranted to provide certain special dispensation to suppliers with sustained exemplary performance related to FM. Such dispensation could include certification status that extremely minimizes incoming inspection and COA reporting quantities. Other bonuses for good suppliers could include being presented the first opportunity to bid on any new business, or products that could be sourced from their facility. No matter how it is done, it is extremely important to recognize supplier locations that are going above and beyond to meet stringent expectations. All too often modern business practices are devoid of this softer skill-set of conveying appreciation for work that is well done. It is good to be a company that suppliers desire to do business with, and to realize that word of draconian supplier management practices will spread quickly among the community.

Possessing a poor reputation could make finding bidders quite difficult for certain commodities that are limited in terms of the supplier base.

2.5 Internal process management systems for FM

Developing specifications, policies, and procedures, and applying source control strategies is the beginning phase of implementing an all-encompassing foreign material control program. Certainly receiving ingredients and product components that are unlikely to contain significant quantities of the FM of concern is a great first step for any processor. The true task at hand, however, is to ensure that the components can be processed and assembled by the processor without introduction of unreasonable FM quantities or types. It is easy to ask a grower or supplier to do the work for you, but in practice the most difficult controls to implement are internal ones. Keeping internal operations in a state of conformity with respect to FM control programs is a task that requires teamwork, diligent systems building, and a high level of process knowledge for understanding the modes of FM introduction. This portion of the chapter will present tools and techniques that can serve as guidance to the processor looking to put FM controls in place, or revamp the existing strategy.

2.5.1 Assessing potential incidents via HACCP and QACCP

Two tools that can be utilized for assessing the potential for introduction of FM into a product in a processing facility are HACCP and QACCP. When conducting a HACCP review, typically the only issues related to public health would be dealt with in the final plan. Much of the FM that is of interest to producers is not necessarily a public health concern. The reader is referred back to the definitions of hazardous versus nonhazardous FM presented in Chapter 1 to illustrate this concept. A product susceptibility analysis must be performed for identification and control of these other relevant contaminants that may impact the overall state of product quality. This review parallels many of the steps used in creating a HACCP plan, but deals only with quality-related issues. A review and management discussion of this type is known as a quality assurance critical control point (QACCP) review for undesired FM.

If a processor does not currently operate under a HACCP plan, then one should be developed and adopted as soon as possible. This text is not meant as an instructional guide to developing HACCP plans—there are many excellent texts that are commercially available as well as guides that are published by various government and regulatory agencies (8 and 9). The HACCP plan should be developed by following either the relevant regulations for the respective industry, or by mirroring the guidelines presented by the Food and Agriculture Organization of the United Nations (FAO). The HACCP preparation document for FAO is quite well written, and an electronic version is provided free to all users via the FAO website located at http://www.fao.org/docrep/W8088E/W8088E00.htm as of the print date.

2.5.1.1 Developing a QACCP plan—Step 1A: gather existing resources

To avoid confusion as well as for legal and regulatory reasons, a QACCP plan should not contain any reference to FM contaminants that could represent a public health hazard. Those types of contaminants must remain referenced in HACCP plans. This will ensure that distinct emphasis on protecting public health is maintained, and should also help to not confuse employees responsible for monitoring HACCP critical control points.

If a HACCP plan already exists for the processing location, the development of a QACCP plan should take much less time. In situations where a HACCP plan has been issued and is in a living state, many components can be reused. The product descriptions, ingredient descriptions, and process flow charts can be shamelessly borrowed from the HACCP plan to facilitate rapid development of the QACCP plan. Before adopting this strategy the processor must first ensure that the HACCP plan is current and accurately reflects the actual manufacturing process used.

Another time saver would be to gather and review any forms and electronic databases where process monitoring information is stored. When working in processing locations that do not practice document control, this exercise can be quite an enlightening experience. It is amazing how much nonproductive work can be discovered and later eliminated when reviewing all of the paperwork that gets filled out in a typical food processing operation. Many times defunct process monitoring records, established by previous management regimes, will continue to be filled out for several years after the last document was ever reviewed by anyone. Make note of any situations uncovered where employees are actively recording nonessential information or duplicating effort by recording information that is already being captured in other areas. When it comes time to implement active process monitoring activities in these areas, the useless information-gathering can be directly replaced by the new value-added work.

Any applicable policies, procedures, and work instructions that address FM in any way should be gathered at this point as well. If the processing operation is part of a larger company, it would be prudent to call around to other locations to identify what policies, procedures, and work instructions they are currently utilizing. In some instances the processor may find that they do not have the most current revision of these materials, and a quick course correction can be taken before continuing forward.

The final piece of data that will be necessary is consumer complaint or comment data for the processing facility. Obtaining summary data over the last three years is a good start. This should help to identify any seasonal patterns that may affect the data, and also allows for one to anticipate rough periods that seasonal operations may experience. Ideally, raw consumer data should be gathered at this point as well. Raw data with verbatim summaries provides outstanding insight into the exact extent and nature of FM issues the consumer faces. Having a good relationship with the internal personnel in consumer affairs or a contact at the agency that collects consumer correspondence can be a big help in procuring this data. Finally, any exhibits collected from consumer/customer returns that have been retained should be gathered for later review by the QACCP team.

2.5.1.2 Step 1B—assemble a cross-functional team

Assembling a team concurrently with the information-gathering process is a wise move from a division of labor standpoint. The best teams for QACCP development usually consist of representatives from the functional areas of:

1. Operations
2. Facilities Engineering
3. Maintenance
4. Consumer Affairs (if available or if it exists)
5. Quality Assurance

This group needs to be hands-on in nature, and possess intimate knowledge of their functional area. Hourly employees with in-depth day-to-day experience should be included on the HACCP team because of their invaluable knowledge of the "real world." Notice that product safety specialists and other finite technical functions typically required for development of a HACCP plan are not required members of the QACCP structure, but these groups are ancillary to this process and should be consulted on an as-needed basis. Obviously they could be included in the group as part of a developmental opportunity, but their time is best spent attending to weightier matters involving public health.

Once assembled, the group should be introduced to the importance of the task at hand and the cost analysis presented to management. The group needs to be formally introduced to the QACCP concept and how it differs from HACCP so that work meetings will not be bogged down. Obviously there are a myriad of other logistical practices that will need to be covered that differ in each processing location. Meeting frequency, times, dates, and locations must be solidified in the first gathering to ensure future attendance of the members. Once the group has reached conceptual alignment with the overall goal and schedule for completion, the real work can begin.

2.5.1.3 Step 2—identify the scope of the plan

The team should identify what the scope of the QACCP plan is going to entail. This exercise consists of describing the products and processing areas that will apply to the plan. In certain situations the development of a formal QACCP plan may only be necessary for various historically troublesome products. Other products may have been produced over the years within the facility and possess an excellent track record for not containing FM. The review of these products may be given a lower priority to address other known issues, but forgetting them altogether is not advisable.

Upon examination of these "model" products, best practices offering quick resolution to the FM problem could be discovered that could rapidly be systematized and reapplied to the problem areas. Conversely, the excellent results observed with other product lines may not be due to sustainable systems at all. If the preventative measures in use for the "good" areas of the operation have not been properly systematized, FM issues may abound as soon as trained personnel leave the area, suppliers change, or new equipment is introduced. This scenario quickly leads to the firefighting mode that was discussed earlier in the chapter. Therefore, review of all product lines while there is team momentum is usually the best strategy.

2.5.1.4 Step 3—identify and describe the relevant ingredients, packaging materials, and processing equipment

Once the detailed scope has been completed, the next portion of the QACCP plan development phase can be tedious. The team must assemble a document that thoroughly describes the relevant ingredients, packaging materials, and processing equipment used to manufacture the products listed within the scope. This documentation will be relatively easy to assemble if the HACCP plan was up to date, but still should be discussed so that everyone is on equal conceptual footing. Each ingredient used should be thoroughly described by its composition, supplier, packaging materials, mode of transport to the facility, and where it is stored once received. The path that the ingredients take to enter the production areas and where they will be used should also be carefully documented by the team. This is important, because in future steps the team will use this information to identify any unsuitable cross-traffic with pedestrians, vehicles, or trash-hauling patterns that could introduce FM.

The packaging materials should be subjected to the same analysis as the raw materials, and documented accordingly. At this time, a revisit of the incoming sampling processes that are relevant to identifying FM in the supplied materials should be reviewed and documented as well. Additionally, traceability measures should be verified, and the manner in which lots are uniquely identifiable can be referenced at this stage as well. Sometimes huge loopholes in purchasing and ingredient segregation practices can be discovered during this stage of plan development. Poor practices typically uncovered are inordinate lot sizes, total lack of traceability for certain materials, and no incoming verification of material quality. Such poor practices would typically have gone unchanged until a problem occurred, potentially resulting in more undesirable firefighting.

The path that the materials follow on their tortuous journey to becoming the finished product should be outlined as a process flow diagram (PFD). The flow path would include pumping stations, conveyors, in-process storage zones (such as heating or cooling rooms), and staging areas for ingredients. A simple block diagram representation of the overall process will usually suffice, and it should include any separation, inspection (both human and automated), and detection steps that are currently part of the process. Any off-line inspection activities for FM, such as analysis of incubation or retain samples, should also be included in the diagram via use of dashed arrows to indicate its off-line status.

A separate document detailing what each step of the diagram represents and what exactly is occurring is also needed to accompany the PFD. The descriptions can be quite brief, but they should correlate to the common name of the equipment, or process, as it is known throughout the facility. A section of driven accumulation conveyor used to deliver empty tin cans to a piston filler may, for instance, be affectionately known as the "can corral." It can be referenced as such in the QACCP plan as long as its function and construction are adequately described. Remember, this is an internal plan that is to be used as a tool, not something that would readily be shared with a third party on any routine basis.

Once the PFD and process description have been assembled, it is a worthwhile activity for the entire team to go out into the operation and verify that it is factual. The path of the ingredients and products should be followed with the following focus priorities (see Figure 2.6):

1. Look above the flow of the exposed product for potential contributors
2. Look adjacent to the flow of the exposed product for potential contributors
3. Look for transient potentials such as sampling equipment and other utensils that would be used intermittently (pens, scoops, calculators, pH meters, etc.)

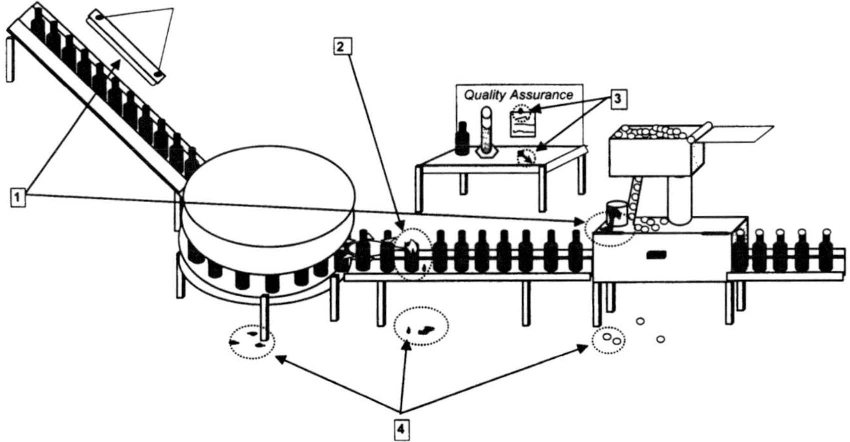

Figure 2.6. Focus points for a QACCP auditor looking for foreign material pathways in order of significance.

4. Examine the area underneath the product flow for potential loose or tramp materials (swarf, concrete, dirt, mold, pallet chunks, nuts, bolts, crimped wires, etc.)

The team members should document any additional threats uncovered during the focused walkthrough with the corresponding position number (1 to 4), and present them at the next meeting. During this floor time the team members should also inspect and observe the operation of any detection or sorting processes to verify that they are still active. One can encounter some rather interesting scenarios during this walkthrough. In past walkthroughs by QACCP teams, situations have been encountered where only the shells of detection or separation equipment remain. In one instance a metal detector was referenced in the flow diagram, and was housed on a process line, but it was completely devoid of a functioning control unit. The most baffling part of this situation was that the operators on the line still referred to it in a manner that would make you believe it was functioning, even though the detector was as useful as a plastic box wrapped around the conveyor.

Personnel practices and traffic patterns need to be examined during this period as well to ensure assumptions made about GMP compliance are still valid. Gross deviations noted in an initial walkthrough can illustrate other areas that may need to be addressed in concert with the QACCP effort.

2.5.1.5 List all potential sources of FM

Each ingredient and processing step listed in the PFD and process description, now confirmed to be accurate, must be scrutinized as a potential contributor of undesired FM. Teaching the group how to use and interpret the output from generally accepted risk assessment modeling tools can be a good first step before beginning this process. Tools such as threat exposure assessment matrix (TEAM) or failure mode effects analysis (FMEA) will help guide the conversation to an endpoint that produces a tangible document that ranks each process step and likelihood of occurrence. The benefit of using the FMEA strategy, however, is that it does provide a means for including the effectiveness of controls already present for isolating the potential FM into the overall risk equation (13). For some reason the food industry has been slow to adopt the TEAM or FMEA approaches even though most governments and other industries have been successfully using them for quite some time.

Regardless of what tool is utilized for quantifying the risk, the type of FM potentially present, the amount likely to be present, and the likelihood of the FM being generated must all be documented. Be prepared for a lot of debate over the likelihood and extent of FM contamination generated as well as the effectiveness of these control measures. To better facilitate these discussions, always ask for a historical reference or any other form of data that could help end the debate over these topics. Each step of the process should be ranked by the team members using the grid presented in Figure 2.7. Remember that the goal of this exercise, however, is not to generate columns of meaningless severity and likelihood numbers, but to identify the group consensus about the true state of risk imposed at each step of the process.

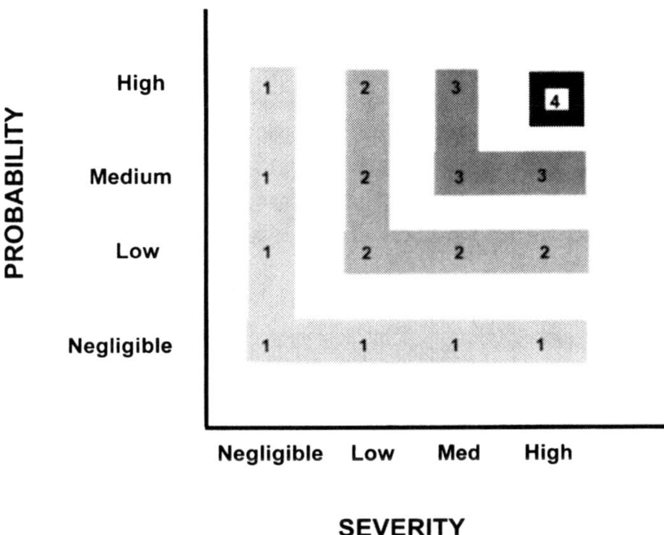

Significance of the Hazard
1) Satisfactory (Negligible)
2) Minor
3) Major
4) Critical

Figure 2.7. Risk classification model for classifying potential foreign material sources by risk.

2.5.1.6 Using output from the risk assessment to identify control points

Hopefully the risk assessment will end with the group having a new perspective about what the identified relative risks of FM are versus their previously perceived risks. Usually several areas will become evident where control mechanisms are lacking or not present at all. To ensure that no stone is left unturned these "Aha" points should be re-evaluated to ensure that a control, such as a COA review or a receiving separation technique, was not missed during the initial review process. If at this point the control factor for the FM is still not evident, it should be noted as a legitimate concern and a potential control mechanism will need to be identified (management practices, sourcing, separation equipment, detection equipment, or other inspection).

The other situation that should be discussed by the team is that of identifying high-risk points that currently list a functioning and adequate control as being present. The question must be asked, "What makes the control point an assured function? How does one identify that this point is working as designed?" Just like a HACCP critical control point, the QACCP team must identify the monitoring technique necessary to ensure that the control is working as designed, and a critical limit for defining satisfactory operation should be stated as well. The notion of frequency for verifying the control's critical limit must also be discussed and agreed upon.

2.6 Error-proofing systems for sustainable results

The effort taken to develop sophisticated foreign material control systems can be undermined if people are left to find their way out of troublesome situations. Development of standardized and systematic troubleshooting documentation, such as decision trees and/or work instructions for operations personnel, can not only aid in the protection of product, but can help to increase productivity. This is especially true when dealing with foreign material incidents that occur at varying frequencies.

When the issue at hand is one that is not encountered on a regular basis, it is difficult for personnel within operations to remember how to proceed in the appropriate manner. Many occurrences will result in an unfortunate situation where the production line will continue to run in the absence of documentation that explicitly states to shut down the process.

When dealing with contaminants that could adulterate the product or otherwise render it unsalable, this is a horrendous mistake that can be financially devastating for both small and large processors. The procedures and policies in place must spell out when to cease operation.

The flip side of this requirement is that, many times, merely giving instructions of when to cease production often results in a process line being shut down for long periods of time for insignificant issues. This time lapse is usually the result of personnel trying to track down more information or awaiting word from the quality assurance or product safety judges who may not be present at 3:00 a.m. This situation can happen even when volumes of policies, procedures, and work instructions are present within a facility because such documents are usually located at the operator's primary decision level. Statements such as "shut down and notify QA supervisor" are typically where most documented quality systems end, giving little guidance to the supervisor for what to do next.

Control plans with decision-making trees should be developed to bridge this procedural gap and ensure that financial viability is maintained while protecting the brand. This is best accomplished by gathering select resources from operations, quality assurance, and accounting to perform an "if then, what to do" optimization analysis. It is best to deal with previously experienced scenarios that were either handled in an exemplary manner or were catastrophic failures as a starting point for this analysis. The accounting group is an excellent resource for providing data on costs that would be avoided when contrasting scenarios and potential solutions. The data from the accounting group is usually a good resource to couple with previous product dispositions for combating the "gambler's" disorder that often affects personnel in production management roles. A typical gambler's scenario would be to continue to produce large quantities of product even though there is a slim chance it might all be subjected to further inspection, rework, or destruction. By framing the circle of events and streamlining the process, a decision tree can be made for the frontline supervision personnel. This newly incorporated institutional knowledge will be based on an optimal business strategy versus tribal knowledge, or whatever the second-shift seasonal supervisor deems an appropriate remedy at the time.

2.7 Quick discussion on product security

Today's environment warrants conducting an FM risk assessment through initial distribution, including malicious tampering concerns. The tamper-evident nature of the packaging function and controls used to ensure limited access of unauthorized personnel to product held under the processor's immediate control is essential. Tamper-proof packaging does not exist, but tamper-evident packaging can help to prevent and/or isolate FM issues that originate from tampering. To prevent confusion on the part of the consumer, the package should give clear reference to and direction for interpretation of the status of any tamper-evident features of the packaging. The reference to the tamper-evident packaging warning should be clearly stated on the products label or permanently coded on the container. What purpose does printing, "Do not consume product if tamper band is missing or broken" on the tamper band itself present? If the band were to have been removed and the product was tampered with, how would the consumer ever know?

2.8 Monitoring tactics to protect the brand after products have been distributed

The last layer of systems necessary for ensuring that FM issues remain under control in the supply chain deal with active monitoring. Many corporations perform one or two of these tasks well, but few are regimented about performing all three. Realize that these activities do not provide 100 percent assurance that all is well in the marketplace, but staying on the proactive side of the continuum can have its benefits.

2.8.1 The voice of the consumer

Another essential activity that helps validate the entire FM control strategy is to vigilantly monitor consumer complaints. So far, the systems developed in this chapter have all been focused on monitoring data from points upstream of the processor in the supply chain, or auditing internal data sources within the processor's manufacturing and distribution facilities to determine the overall system capability. Data derived from the consumer feedback system can serve as a key performance indicator for these other systems utilized within the supply chain. Monitoring consumer complaint data can not only ensure that these systems are working as designed, but indicate whether continuous improvement activities are meeting expectations.

Consumer feedback is truly an indicator of the quality of the product distributed, but mishandled consumer complaint information can become damaging evidence in a product liability case. The information received from the consumer must be reviewed in a timely manner, and follow up is absolutely essential. Realize that by quickly quantifying and reacting to "real" issues in a diligent manner, the scope of potentially large-scale product incidences can many times be contained. Prioritization of complaint handling can be streamlined by assigning a severity index to each complaint category, or comment type. Chapters 9 and 10 of this text outline appropriate monitoring strategies for consumer complaints and the development of systems to react to issues that could be indicative of serious issues.

2.8.2 Go shopping and see what the product really looks like

Do you want to know if that new packaging design survived the customer's warehousing techniques? Did the contract manufacturer really have a good day with those out-of-spec materials they had to use? Why are all these complaints coming in for this specific contaminant in just these three cities? How can the competitor make that product and not have tons of defective units out there? How close are they to that DAL? These questions can usually be effectively answered by performing a store audit, and purchasing some product for analysis.

Examining products that have been fully distributed gives the processor a much-needed sanity check for what the customer and consumer are actually presented with upon a visit to the store. Regional clustering of imbedded insect complaints and other FM are sometimes rooted in the storage practices used at the store or customer distribution centers. Sometimes a shopping trip can reveal product that had been originally dispositioned for donation or destruction by the processor (because of aesthetic FM issues or worse), but is now being sold on the open market and generating complaints!

The data obtained from shopping trips doesn't usually represent a large enough subset of the lots produced by the processor to be considered a statistically significant representation of the population. It is still useful, however, in chasing down specific answers to questions related to FM in foods. For this reason, it is a tool that should be ready for use by the processor whenever questions arise.

One point to recognize, however, is that going shopping no longer requires the use of internal resources. There are several competent third party services currently offering product gathering and retrieval services to food processors at modest fees. The processor just has to provide instructions relevant to the what, where, and when of the desired procurement exercise, and door-to-door service can be expedited for quick results.

2.8.3 Monitoring the mass media and surveillance of the Internet

Even though the increasing network of global communications available have presented outstanding marketing channels and information sinks for corporations to exploit, it has not been without drawbacks. Realize that not all disgruntled customers, consumers, former employees, or special interest groups are going to call the company's 800 number and tell them how they feel. The World Wide Web provides a global billboard for the chronicle and promotion of both legitimate and fraudulent experiences by such groups, and the ease of becoming registered with legitimate search engines makes it a challenge for corporations to address consumer inquiries related to such internet postings. If the websites that are listing these claims are vibrant, appear to be professionally developed, and seem legitimate, many people will attribute credibility to their content.

Though no reputable journalist would promote and publish "mouse in can" stories found on the Internet, in mass media, or in print, many consumers will propagate the stories to others as fact. This type of communication offers many challenges for becoming aware of and containing pseudo-incidents. To ensure that FM issues are not being overblown and underreported, there should be frequent reviews of the Internet and the media for references that implicate the company and its brands. This task can be managed in many ways, but some of the most popular contemporary approaches are to:

1. Identify the top five search engines utilized on the World Wide Web, and perform daily searches using appropriate combinations of your company and brand names with negative operators
2. Contract with an electronic news service and database information aggregator to cover legitimate media (4)
3. Use an internet monitoring and clipping third party service to track other "nonlegitimate" publications on the World Wide Web (4)

All of these scenarios obviously represent an incremental cost to the processor. The costs will be incurred in the form of either the contracted third party or the dedication of time via a finite internal resource. Realize, however, that these costs can become insignificant compared to the ill effects suffered because of an illegitimate claim that becomes "factual widespread knowledge" or a legitimate claim that was not handled well by the company's consumer relations department and thus became the subject of a high-traffic weblog (Blog).

Works Cited and Select Resources

1. Brett J. 2002. *You Can Prevent Hardware Disease.* Dairy Herd Management (7):35.

2. *Force Field Analysis.* 2001. Ames, Iowa: Iowa State Univ. Extension. Available from: <http://www.extension.iastate.edu/communities/tools/forcefield.html.> Accessed Apr 13, 2004.

3. Heaton K, Bagley C. 2002. *Foreign Object Contamination in Beef Cattle (AG 516).* Logan, Utah: Utah State Univ. Extension. P 1–2.

4. Kassel A. 2000. *Guide to Internet Monitoring and Clipping.* Available from: <www.cyberalert.com>. Accessed May 22, 2004.

5. Moseley BL. 1999. *Hardware Disease of Cattle.* Agricultural Publication G7700. Columbia, Mo.: Univ. of Missouri Extension. P 2.

6. Franklin A, Neva FA, Brown HW. 1994. *Basic Clinical Parasitology. 6th ed.* Stamford, Conn.: Appleton & Lange. 650 p.

7. Pehanich M. 2003. *Tech Update: Powder & Bulk Handling Equipment.* Food Engineering (8). Available from: <www.foodengineeringmag.com>. Accessed Jan 7, 2004.

8. Shapton DA, Shapton NF. 1991. *Principles and Practices for the Safe Processing of Foods.* Burlington, Mass.: Butterworth-Heinemann, Ltd. 457 p.

9. Stevenson KE, Bernard DT. 1995. *HACCP—Establishing Hazard Analysis Critical Control Point Programs. 2nd ed.* Washington, D.C.: The Food Processors Institute. 224 p.

10. Snyder JE. 2002. *Corrective and Preventative Action: Planning to Achieve Sustainable GMP Compliance.* J of GXP Compliance 6(3):29–39.

11. Van Donkersgoed J. 1999. *Broken Needles in Beef.* Calgary, AB, Canada: Canadian Cattlemen's Association. P 1.

12. Vansickle J. 2001. *Broken Needles: One is Too Many.* National Hog Farmer (1):22–24.

13. *Procedures for Performing a Failure Mode, Effects and Criticality Analysis.* 1980. Military Standard 1629A. Washington, D.C.: U.S. Dept of Defense. 54 p.

CHAPTER 3

Physical Separation Techniques for Controlling FM Contaminants

3.1 What is the difference? Physical characteristics of product versus the FM

Sometimes developing separation schemes for FM is as simple as understanding the subtle differences that exist between your product and the contaminant(s) of concern. Simple questions can often illustrate this concept quite well—(1) Does the product float? (2) What are the typical dimensions of the product? (3) Does the product break easily? (4) Would the product stay suspended in an air stream of moderate strength? (5) What is the bulk density of the product versus the contaminant of concern? (6) Does the product flow easily under the influence of gravity?

Most corporations have characterized the physical nature of their products well after many years of manufacturing them. Typically such physical attributes to answer the above questions are monitored, and recorded in a high relative frequency by personnel as key measures for ensuring product quality. By mining this data and performing confirmation studies, a great deal of insight can be gained on how one could use the theory of a subtle physical difference to initiate a separation strategy. The purpose of this chapter is to introduce the reader to the many techniques that are used throughout the industry to exploit these sometimes subtle physical differences.

This chapter is essentially split into general segments where differing types of separation techniques are presented. The first half of the chapter deals with separation techniques used mostly on dry food products, while a large portion of the second half is dedicated to separation techniques that can be used on wet food products. Suggestions for process monitoring, creation of records, and tying separation strategies into existing HACCP and quality plans will be covered in the summary section of the chapter.

3.2 Sifters and screeners for free-flowing dry products

The task of employing physical separation in dry products is aided mostly by the size and shape of the contaminant versus that of the product. One method that can be used to determine any differences in these physical characteristics is to sift or screen the product. Essentially, this involves allowing the product to be exposed to the surface of a screen whose size is determined by the nature of the product versus the contaminant. If a screener were to be used at a dry ingredients bag dumper, for instance, the foreign material of concern may be rogue pieces of paper that could measure anywhere between 1 to 10 cm. If the product is granular in nature, it is easy to see that the screen could be sized in a manner that would not impede the flow of the ingredient, yet ensure that the contaminant would effectively be blocked from admission into the product.

Obviously the example illustrated above is very simplistic, but hopefully the point is clear. Selecting the appropriate screening equipment for an operation is many times a task that requires thorough knowledge about the characteristics of the material inputs and the needs of the process. Some materials cannot simply be poured through a screen at a high rate of speed due to their particle size or other characteristics such as moisture, shape, density, electrostatic charges, etc. These situations call for unique designs that use varying techniques to present the material to the screen in a continuous manner. There are essentially four different designs of screeners commercially available for use, and they are classified mostly by the forces they generate and the motion of the material relative to the surface of the screener. Realize that no matter which screener is chosen for application in a processing environment, there are only two relevant modes of separating FM from the product. Both of these may occur concurrently if the proper design is chosen. The two modes used to separate FM from a food product using a screener are:

1. Scalping FM from the material stream that is greater in size than the product, or whose shape prevents passage through the screen
2. Removing FM that is finer than the product and is retained in the undersized fines stream

These principles are widely applicable and true of other screening-type operations such as roller-operated sizers used to grade produce items.

3.2.1 Materials used in the manufacture of screens and their sizes

The most important part of any screener in preventing FM is the screen itself. The screen in a separator can be composed of different types of materials. The two general categories of material types that can be used to manufacture separator screens are metals (stainless and carbon steel) and synthetic polymer strands (nylon, polyester, etc.) (4). Essentially the benefits of using a synthetic mesh are that they are not prone to corrosion or reactivity with the product matrix, and additionally they are not prone to harden over extended periods of use. The advantages of metallic screens are that they generally provide good strength characteristics at smaller diameters than their polymer-based counterparts (4).

Experience dictates that all screens will fail, given time. The mechanisms of failure vary, but most can be categorized as occurring through either:

1. Defects in the manufacturing process of the screen
2. Misuse at the facility or improper installations
3. Wear or catastrophic unique events (1) (think Murphy's Law here; i.e., dropped tools, wandering operators, etc.)

When any of these events occur, the screen material can easily become FM itself. For this reason, a screen made from magnetic highly conductive material that is readily isolated by a detector (such as 400 series stainless steel) and prone to magnetic separation techniques is preferable.

3.2.2 Placement of screens within the process

When dealing with dry products there are three generic processing points where installation of a screening system could prove useful in ensuring that FM issues do not proliferate. The first location would be screening upon the receipt of materials from bulk sources such as rail cars, trucks, or totes as they are in the process of being transferred to surge bins or silos. Not only does screening at this step provide assurance that certain undesirable FM are not contaminating existing inventories, but it can provide for instant dockage of suppliers or transporters that allow contamination to be shipped to the facility.

The second location where screening can be applied is at the input stage of ingredients or in-process materials into subsequent process steps. Bags of flour, pasta, or flavoring agents can usually be quickly sifted to ensure that "reasonably likely to occur" contaminants aren't present. By performing this step in a routine fashion, one can also be assured that tools, utensils, testing equipment, and personal effects of those in the vicinity do not enter the process as materials are manually transferred or dumped.

The final location for a screening operation is immediately prior to the filling operation. Many times this is an "oversize" operation to ensure that the product is liberated of any incidental contaminants prior to being delivered to the carton, bag, can, or jar.

3.2.3 Types of screener applications available for use with dry products

There are four types of screeners commercially available for dry product separation within the food industry. The general categories are:

1. Centrifugal
2. Vibratory
3. Gyratory-Reciprocal
4. Gyratory

Examples of each type and factors governing their use will be discussed in the subsections that follow.

3.2.3.1 Centrifugal screening separators

Generally centrifugal screeners are not considered a compatible design for a vast majority of food processors (according to some baking industry standards), due to nuances dealing with their functional design (4). They can serve the purpose of sorting out potential foreign material contaminants quite well, however, as long as the type of contaminant is not readily friable (easily fractured into smaller pieces) by the mechanical action applied. Centrifugal screeners can also be used for special purposes when trying to quickly deal with the sorting of lots that are suspected to be contaminated. Essentially, a centrifugal screener is designed in a format where the screen (of fixed size) is mounted in a fixed fashion and fed via a screw auger. The feed meets a rapidly rotating paddle or series of beater bars encased in the center of the screener, and it is forcefully launched toward the waiting screen for classification. Oversized products that do not break apart will not pass through the screen and be retained in the housing.

Obviously only certain types of products are appropriate for this type of separation due to the violent nature of the process. The screen would need to be designed for a combination of either shape selection for the product (round, square, etc.) or size selection for the product. The configuration would then allow the product to pass through the screen, leaving the foreign material to cascade to the collection area at the bottom of the unit. In a dry product environment this would be very effective for liberating contaminants like wood, certain plastics, and soft thin metals like aluminum cans which are not yet of a size that would pass through the screen. Obviously this type of screener would be a poor choice for attempting to segregate glass or hard/jagged metal or plastics. Drawbacks to this type of design are: it could require a great deal of energy to continuously power the spinning paddles, and the constant impact of the product with the screen could cause wear through flexing. Wear of this type usually will result in the screen breaking at some point, which could then become a source of future contamination.

3.2.3.2 Vibratory screening separators

The use of vibratory screening separators is quite common in dry product handling applications. This type of separator yields good separation of contaminants that are of a different dimension from the product. A vibratory separator is essentially a series of round horizontal screens that are stacked vertically within a frame assembly (see Figure 3.1). The screen sizes are chosen based on the particle size characteristics of the desired product, and the screens will be stacked, from top to bottom, by decreasing-sized holes. In most applications each screen has a vertical post through its geometric center that fits into a welded bushing and consists of a threaded bolt possessing a washer and nut. The tension of the screen is adjusted via tightening or loosening of the nut—a critical maintenance activity that, if not done properly, could result in premature screen failure.

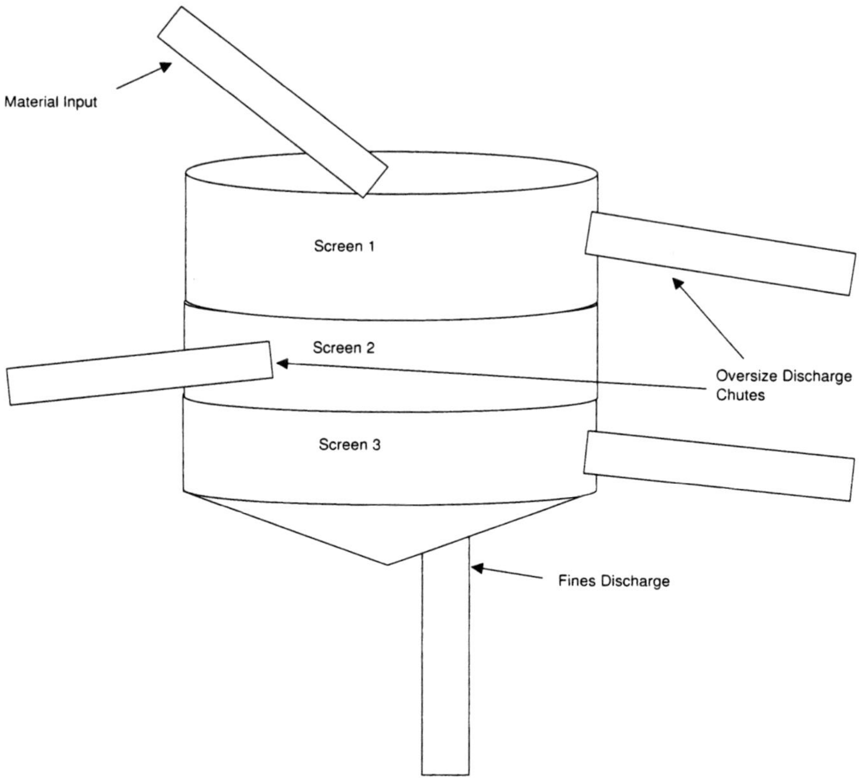

Figure 3.1. Typical vibratory separator.

The motion applied to the screener originates from an eccentric drive. Material is fed into the top of the separator, and the eccentric drive provides for a series of motions that are short in interval, back and forth (on a linear plane), and coupled with vertical motion (4). Each vertical screen location has an outlet port that the oversized particles will exit through. Material that is able to pass through the screen will drop to the next smaller screen size housed below. Once there, further separation, or pass through to the bottom exit, will occur if selection is complete. This design is ideal for screening out both oversized FM (nuts, bolts, rocks, wood, stems, etc.) and undersized contaminants from dry products (sand, grains of metal, small glass fragments, etc.). Another positive outcome from using a vibratory separator is that the motion of the separator is many times not intense enough to fracture friable FM like glass, metals, and hard plastics. One drawback of the vibratory separator, however, is that it will not break up products that may clump together because of moisture or oil content, so its application needs to be carefully examined. Additionally, it is difficult to get high throughputs from the units due to the inefficient use of the screen area. Essentially the process throughput will be bottlenecked by the throughput of the area of screen sized to endure the brunt of the material selection.

3.2.3.3 Gyratory-reciprocal screening separators

The construction and operation of a reciprocal separator is actually quite simple. The separator is composed of a rectangular screen that is mounted in a frame with side walls. The screen is set up in a horizontal (lengthwise) manner, but positioned at a steep angle. The drive mechanism of the separator is such that it creates a gyratory motion at the inlet of the sifter and a reciprocating linear motion at the outlet of the separator. The net result of these two induced forces is a gentle elliptical motion that helps convey the material across the surface of the screen while providing motion for selection.

The product will either be pneumatically conveyed or gravity-fed to the elevated end of the separator. Then the material will either fall through the screen or reach the outlet at the low side of the separator. The sloped design of the separator and its elliptical motion provide for high transfer rates through the screener. One drawback of the gyratory-reciprocal separator, however, is that it only possesses a single screen, so careful planning for screening out FM contaminants is essential. A second drawback to this separator design is that it requires a rather large physical footprint for installation in an existing plant.

3.2.3.4 Gyratory screening separators

A gyratory screening separator consists of multiple horizontally mounted screen decks that can be square or round in shape. The horizontal relationship of the screen decks is similar to the vibratory screener: however, the material handling and motion of the separator are different. The screener only possesses horizontal gyratory movement. This makes for a very gentle motion that will not break up clumps or agglomerates.

3.3 Use of gravity separators with dry products

A force that governs all food products produced on this planet is gravity. Not all products will be a candidate for gravity separation, but the technique will be covered here to present the option. Gravity separators, commonly referred to as air tables, can be used in a continuous or batch process to eliminate undesirable components from food products when certain criteria are met. Depending on the design of the gravity separator, a contaminant that is lighter than the desired material (such as cellophane in flaked cereal) may be removed or, in the case of a stone in the same flaked cereal, a heavier fraction may be removed from the product stream.

The product must have a rather high flowability as measured by a low angle of repose (<60°) to be a good candidate for gravity separation (4, 3) and the bulk density of the material must differ consistently from the contaminant that is to be separated. It is also desirable that the material to be separated possesses a rather uniform particle size (i.e., as a rule of thumb the smallest acceptable particle should be no less than half the diameter of the largest particle). Finally, the material will need to have a uniform aerodynamic shape so it will react similarly with the air currents introduced. If the product meets all these conditions, it can be efficiently separated from another material of differing characteristics (contaminants).

The motive forces of a gravity separator are provided by an eccentric vibrating drive usually operating between 200 to 600 Hz. The eccentric drive provides a side-to-side motion, and a strong fan provides vertical air currents. The separator is usually constructed with a rectangular wire screen that is sized to allow sufficient airflow, but will not allow the product to pass through it. The screen will have multiple solid guide strips, called riffles, mounted in an offset manner across its surface. The screen itself will be mounted on a lengthwise slope with an adjustable pitch angle across its width (3). This setup resembles a pinball machine that is tilted on one side toward an angle. A tilt of this magnitude in a pinball game would certainly result in a loss, but when playing the game of separating foreign materials from food products this tilt is an essential component of the separator.

With the fan and drive engaged, the material to be separated is fed to the high side of the inclined deck. The feed rate is usually modulated at this point with a slide gate to ensure that the deck is not overloaded with material. As the material falls onto the deck, it meets the vertical force of the air currents and the horizontal force of the drive. The lighter fraction of the material will fluidize and ride on top of the heavier fraction and will drift to the low side of the tilted platform under the influence of gravity. The heavier fractions, however, aren't readily fluidized by the air currents and will travel uphill, due to the lateral motion exerted by the drive. The uphill side of the screen possesses an adjustable grate that the heavy fraction will pass through and drop off the edge of the deck. Likewise the lighter fraction will head to the lower side of the deck and be discharged through a different chute.

Gravity separation is a useful technique, but if your material is highly variable then its applications are somewhat limited and use of the technique could be quite costly due to high reject rates. The separators require energy for the fans and the eccentric drives, and the air supplied to the fans should be appropriate for direct product contact. Installation of a gravity separator will require space, and in many cases an operator will have to be stationed in the vicinity to ensure that it is running properly (no plug-ups or overflow situations).

3.4 Air classifiers

Air classifiers are widely used in the grain industry for separating undesirable components such as chaff or brokens from good grades of material. In any air classifying system that will be used to separate undesirable contaminants, there are two predominant forces that are used to exploit varying product characteristics. These forces are aerodynamic drag force of the material and a mass force present in the gravitational or centrifugal field (3). The function of any air classifier is governed by complex mathematical formulas related to the motion of solid particles when external forces are applied. The math can get quite confusing because the equations used to describe the motion of particles under forces differ as the size and mass of the particles change. To prevent confusion and for the sake of brevity, only the types of commercially available air classifiers typically used will be presented conceptually within this text.

3.4.1 Gravity air classifier

The crudest application of air classification is the gravity classifier. This application is composed of a vertical zigzag channel where an air current is introduced toward the bottom. The material to be classified is introduced approximately in the middle of the channel, and the fines are eluted by the air stream through an exit in the top and into a collection unit. The heavier mass fraction of the material (potentially a stone or nail in a wheat stream) will succumb to the forces of gravity and fall to the bottom. This type of setup has advantages over traditional sieving because it can separate by density versus size and shape of the particle.

3.4.2 Cyclone separator

Cyclone separators are quite useful for the separation or concentration of fine material from fractions that may contain undesirable contaminants (sand, metal fragments, etc.). The cyclone concept can also be used for sedimentation of impurities and contaminants from liquid products. Material is introduced tangentially to the side wall, and there is a fines outlet at the top of the cyclone that will have either a classification wheel or a vortex finder (a piece of pipe that extends well below the inlet to prevent materials from short-circuiting to the outlet) associated with it. The air and product mixture will need to be introduced at a sufficient velocity to cause a vortex. Denser particles will usually impact the side of the separator due to their inertia, and will be outside of the high-pressure region of the vortex. Diagram of a typical cyclone separator is presented in Figure 3.2.

Due to the lower pressure gradient exerted externally to the vortex, the denser particles will fall out of the air particle mixture and be taken by gravity to the discharge. Lighter particles will maintain their velocity when entering the high-pressure region of the vortex and follow the air currents to exit the top of the cyclone. The design of the cyclone is critical for its operation. The diameter, taper, and vortex finder depth of the cyclone are all critical design factors for determining what size, and density, of fine material will be able to travel to the outlet. Additionally, if a selecting wheel is to be used its design must be subjected to the same level of scrutiny as the factors above to ensure that the separator will function as desired.

3.4.3 Combination elutriation-vortex separator

This classification scheme uses centrifugal force to separate larger particles from those that can be drawn in an air stream. The product that is already flowing in an air stream is introduced to a chamber where it meets

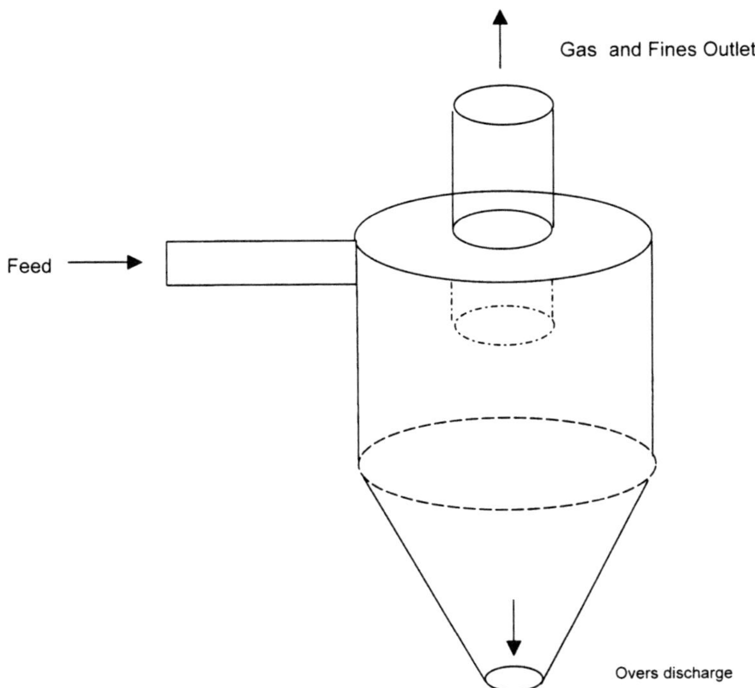

Figure 3.2. Schematic diagram of a cyclone showing the flow pattern.

the tip of a cone positioned at the inlet. A rotating classifying wheel is at the base of the cone that helps to create a forced vortex of the product air mixture. The cone and wheel setup are mounted directly below the fines outlet of the separator. The inertia of the coarser particles will be influenced by centrifugal forces imparted by impact with the wheel, and they will travel toward the periphery of the chamber. Here they will fall to an outlet for collection. The finer particle fraction will follow the air stream and exit the separator through the fines outlet. Typical operation is presented in Figure 3.3.

3.5 Addressing the separation of contaminants from wet product streams

This section will deal with separation of foreign material from products or ingredients that are wet in nature. "Wet" will be defined hereforth as

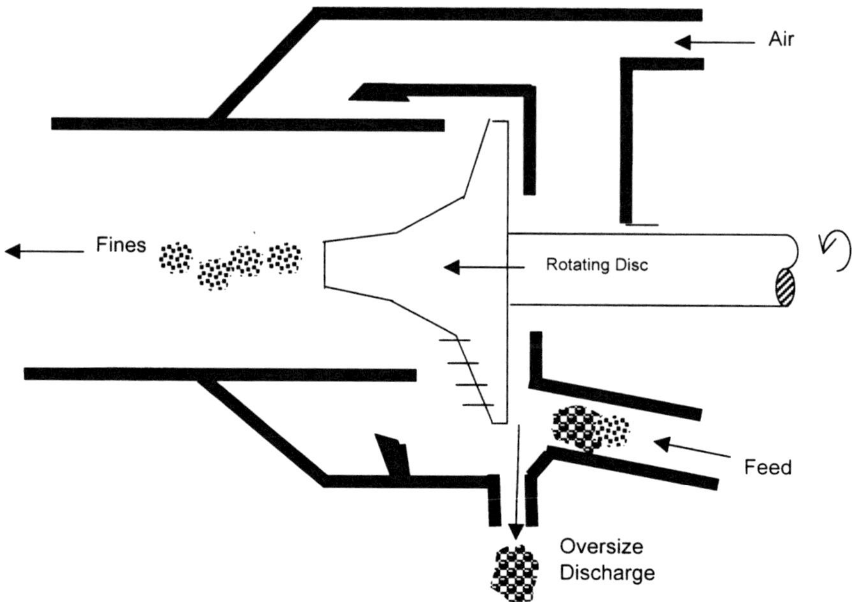

Figure 3.3. Mode of operation of a centrifugal air classifier.

products or ingredients that are measurably composed, in part or wholly, of an unbound solvent such as water, oil, or an organic carrier solution (vinegar, ethanol, etc.). Wet products can be solid or liquid, but all are susceptible to FM contamination if preventative measures are not taken.

3.5.1 Gross screening of wet products

Screening wet products such as fruit and vegetables is a common practice. The product can be conveyed over a coarse screen in manners similar to those described for dry products. The end result will be the gross separation of residual unattached plant material and nonimbedded contaminants from the produce stream (such as rocks, nails, screws, and glass).

A dry trommel screen may be used for applications where the material is either not prone to damage, or where it will be peeled, sliced, ground, cut, or pureed at a subsequent step. Trommel screens are essentially large rotating horizontal drum screens that are powered by external drive mechanisms. Figure 3.4 illustrates a typical trommel screen configuration used to physically separate products from undesirable materials. They are of variable speed, and their rotation exposes the product to the screen surface in a repeating manner. The undesirable smaller portion of the material should be mostly eliminated if the screen is sized appropriately in length and hole size. Throughput through the trammel screen is aided by placing it at an angle to facilitate flow, and ensuring that it is turning at a fast enough rate to keep material from sitting stationary (2). When operated in combination with a circulating shallow water bath and centrally mounted spray bar, the screening setup can double as a wash reel for the product. Having a shallow pool of water at the bottom will yield yet another benefit of providing a more subtle landing for product that "takes a ride" up the side of the drum, and can help prevent product damage. If retention times against the screen need to be increased for additional sorting effectiveness, vertical dams can be mounted in the trommel screen. By adding vertical dams of differing heights, the effective residence time of material in the screener will be increased. Realize, however, that taking such measures will reduce the overall efficiency of the screening unit.

From a maintenance standpoint, trammel screens will require periodic attention. They are, in essence, a moving part by design so the drive mechanisms will require lubrication and inspection. Care should be taken to provide for drip pans, and grease fittings that are placed in a manner away from the product flow. The trommel screens should also be inspected periodically for loose or missing fasteners, gouged screen surfaces, and broken screen areas (2).

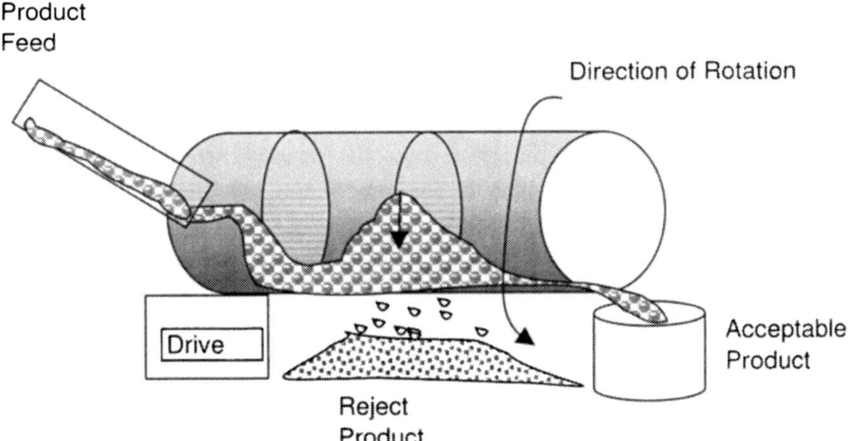

Figure 3.4. Operation of a trommel screen.

3.5.2 Piercing panels (pin conveyors)

To remove fibrous stringy contaminants such as hair, stems, stalks, twine, and loose soft plastic from process streams, more aggressive separation techniques may sometimes be required. Using a vibratory pin conveyor (piercing panel setup) can effectively cull contaminants from produce, confectionary items, and certain types of meat products that otherwise would not become unattached in a normal screening operation. The operation of a vibratory pin conveyor is essentially the same as a screener—except that instead of the items traveling on a screen, there is a peg setup that resembles a bed of nails. The design can also be used to lightly puncture the outer layer of products to allow for venting of gasses during the cooking process or aid in the equilibration of brine solutions introduced later.

Operation of the pin conveyor is relatively simple. Materials are conveyed across this panel, usually up an incline, and the undesirable smaller pliable fraction is either ensnared in the network of pins or impaled by them, ultimately impeding their travel. The product is moved to the next conveying or process operation, leaving the undesirable material behind. Needless to say, cleaning of this type of separator is not an easy (or very safe) activity for the inexperienced sanitation team member. Additionally, enumeration and subsequent documentation of the material separated is usually not possible due to issues with access and cleaning of the separator surface.

3.5.3 Filtration of wet product streams

Actively pumping liquids through strainers and filters is a great way to separate undesired foreign material from the product stream, and at the same time protect expensive process equipment such as pumps, filters, and heat exchangers. Essentially this technique works well for separating FM from products. This is true as long as the product to be strained is absent of particulate matter—juices, oils, dairy products, bullions, fine purees, some soups, and beverages such as soft drinks and beer are all excellent candidates. Essentially any liquid product can be a candidate for this operation as long as it is not overly sheer sensitive or contains particulate of a size and resiliency that would not allow the product to be handled in this manner (such as potato soup).

3.5.3.1 Strainers

A typical straining setup can be achieved via use of coarse or fine low-pressure drop metal strainers to capture contaminants. The product that is to be strained is pumped to a housing that contains the strainer (usually a metal screen). Here the product will be introduced to the inside of the housing where it will contact the exterior of the straining screen and the "good" product will pass through the holes in the metal sieve and flow to the outlet in the housing. Any contaminants larger than the holes in the sieve will be trapped between the strainer and its housing, where they will remain for later removal. The strainer will remain effective until all of its holes are blinded (plugged), which results in a reduced flow of the product. Usually this situation can be avoided by the installation of a pressure gauge or transducer on the inlet side of the strainer. As the strainer becomes blinded, the back pressure on the inlet side will increase substantially, indicating that there is some material blocking the flow and the strainer housing should be disassembled for inspection.

Fewer commercially available strainer setups can be ordered in a continuous multiple strainer configuration. In this mode, the product flow path is diverted at different time periods to another set of strainers and the strainer previously used is backflushed with a high-pressure water rinse to remove any contaminants that may be present on the screen or in the housing. The backwashed particles can be routed to a collection box screen setup so the rejects can be examined for the presence of FM. Other designs that incorporate a self-cleaning feature are set up where the product is pumped into the center of the screen and will flow through the screen unless FM or large particles are present. Periodically a plunger-type disc that is of the same diameter as the internal diameter of the screen will be actuated to scrape the screen sending the oversized materials to a waste outlet.

Regardless of the screener's overall design, it should be inspected regularly to ensure that the screen is intact (no holes) and seated properly in its housing. By not fully seating the strainer in its housing, contaminants could short-circuit the process and travel unabated to the filling area. Routine checks of the strainer can also alert the processor of the presence of any potential contaminants that may be present in the product. If the contaminants are allowed to rest in the screen housing assembly, the product flow may promote collisions of the contaminant with the screen potentially breaking the screen itself or fracturing the foreign material into smaller pieces that may pass through. Operators responsible for screen assembly and disassembly must be educated in the importance of correct assembly, and given documented work instructions on how to react to screen damage and/or FM finds.

3.5.3.2 Sock-style bag filters

Sock filters are similar to strainers in the respect that they are seated within a housing in the product flow path. A sock filter is composed of a synthetic woven material and usually possesses a much finer mesh than a sieve. Sock filters can possess a static diameter for filtration of less than five micrometers, and can effectively be used in dry product environments as well. The sock filter is primarily the last line of defense present in the final clarification of transparent liquid products such as water, juice, or beer to ensure that no "floaters" or visible particulate matter will be present in the finished package. A hard or sharp foreign object may damage the filter bag and rip it open, so they are not an end-all stopgap for all types of foreign material. They do, however, provide an excellent defense for preventing contamination such as chunks of burnt particles (from heat exchangers), gaskets, human hair, trub, large particles of rust, or paint flakes from being retained in the product stream. The surface area of the bag filter is finite, so its effective life of operation depends on the amount of suspended matter in the solution that will be passed through it. Proper installation of the filter bag is essential for it to prevent foreign material from short-circuiting through the seat and contaminating the product. As with a strainer, all finds in the bag filter should be dealt with through execution of written procedures. Bag filters should also be changed frequently according to the manufacturer's recommendations to ensure they do not fail unknowingly while in use.

3.5.3.3 Depth filters

Depth filters (commonly called cartridge filters), when they are appropriately sized and contain the correct selection media, are used to filter unwanted biological and chemical contaminants as well as FM from media used in food processing. Depth filters can be used to clean air or other gasses that come into contact with products, process water, steam, and the liquid product itself. The media in a depth filter can vary depending on application (carbon or ionic resin, for instance), but the action on FM for all depth-filtration systems is consistently based on physical selection by size and shape. Due to the fact that it can only hold so much material within its structure before flow is prevented, a depth filter has a finite service life and will need to be changed periodically. Much like strainers and filter bags, the depth filter is usually contained in a cartridge housing that must be properly seated to provide adequate filtration.

3.5.3.4 Plate in frame filters

Plate in frame filter configurations are used to batch press and extract juices from a mash of trub and other impurities (as in the case of grape pressings used to make wine). Multiple screens will be stacked by descending hole-size in metal frames that move vertically. The frames are tightened against each other via a hydraulic ram to ensure that liquid does not leak out of the side of the press once it is introduced. The product to be purified is pumped into the plate housing and is presented to the screens for filtration. Hydraulic pressure is applied to the assembly, and the clarified liquid exits through the fine screen. The assembly is opened when the product ceases to flow out the outlet, and the remaining solids that form on each screen (known as filter cake) are cleaned and disposed of. Plate in frame filters are very effective at removing foreign material the size of yeast cells from the liquid product streams, and are typically operated as part of a slightly antiquated juice pressing or winemaking process.

3.5.3.5 Membrane filtration

Membrane filters have been around for quite some time in the food industry, and are usually used for highly selective liquid separation activities to rid products of impurities such as microorganisms and particulate matter finer than a single micron. Juice pressers, wineries, and breweries use membrane filters for ultrafiltration to concentrate desired liquid fractions within their products while leaving the undesirable fractions that contribute to off flavors behind. Another common commercial application of membrane filters is in the production of reverse osmosis-treated waters to render them commercially sterile. Membrane filters work well at preventing foreign material, but the initial investment costs, maintenance requirements, and system-monitoring procedures required to keep a membrane system operational for the sole purpose of preventing FM is a poor investment.

3.5.4 Washing and physical agitation of food products

If the contaminants of concern are not of a jagged shape that would allow them to be imbedded in the product, elutriation of FM via washing and scrubbing is a viable application. This basic technology is used throughout the food industry when products can tolerate a short stint in a water- or other approved-solvent media. The product may be subjected to high pressure water jets to remove dirt and residual FM potentially present in the product stream. Wet brushing of the product via conveying the product over a series of submerged oscillating brushes, or rubber rollers in a shallow tank of water is another method that can be employed to remove small stones, twigs, cardboard, stems, and leaves from the product. There are many commercially available deleafers, circulating water baths, high-pressure water spray booths, and brush washers available to suit a plethora of wet product applications. If a brush washer is to be used, ensure that the bristles of the brush are suitable for the application. They should be made from a material approved for food contact. The material should be highly durable, yet pliable enough not to fracture under the stress of the operation. If the wrong brush material or style is used, the bristles may end up as an undesired contaminant in the product.

3.5.5 Product contaminant density-related buoyancy

When products are conveyed into water, not only are they potentially liberated of contaminants through the washing action described previously, but they may travel away from the contaminant via density separation in the water medium to prevent further commingling. This category of separation relies on whether or not a product will sink or float in a different manner than the FM contaminants of concern. This is a very simple concept based on the density of the material versus the contaminant that is often overlooked in the design of many processing lines for fruits, vegetables, meats, and poultry. Contaminants with a high density such as glass, metal, stones, and hard plastic will sink over time when placed in water. Other contaminants such as wood splinters, chunks of pallets, stems, paper, cardboard, and foam may float on the surface of the water depending on their respective density.

Float tanks that use guided water jets to push the product across the surface of the tank or along the bottom of the tank are a typical application used in fruit and vegetable processing to exploit this stratification effect. By strategically placing dams, or weirs, in series, the tanks are sectioned to provide for progressively cleaner fractions as the product travels to the next processing area. The tanks can also be constructed with horizontal screens suspended in the water to isolate different fractions of material based on density, and can weed out defects like bloaters, split cores, etc.

3.6 Using physical separation techniques to eliminate contaminants from packaging materials

It is well established that imbedded or loose FM can be present in food packaging due to the manufacture, transport, and storage methods used by suppliers, haulers, and processors. Many separation techniques exist for dealing with loose foreign material that may be present in rigid and semi-flexible containers manufactured from glass, metal, and/or extruded plastics. Again, the concept that the foreign material is inherently different from the container can be used to liberate the contaminants from the container before they become a problem.

The public health significance of imbedded contaminants such as "stuck glass" or can flashing can be debated. They are not, however, usually effectively separated from the container via physical separation methods. Detection strategies presented in later chapters will be more suitable for dealing with imbedded contaminants.

3.6.1 Inversion wash

The most effective way to ensure that loose product is removed from containers is to use an inversion wash step. For this operation the containers are usually conveyed to the filling line in mass, and introduced to a single file conveyor arrangement. From this point a few feet of conveyor is present (not necessarily the case with cans, as illustrated below) to ensure adequate back pressure is placed on the containers as they enter a twisting inversion rail. With the container now upside down, a blast of high-pressure air or water is introduced to wash away any contaminants that may be present. In the case of plastic bottles, a deionizer should be used on the container prior to the high-pressure blow to ensure the contaminants are not held to the container's surface through static-induced attraction. The pressurized rinse setup within a container inverter is preferred rather than relying on gravity alone, due to the nature of the contaminants that may be present in the container (hair, dust, personal effects, glass, swarf, etc.).

For metal cans, a magnetic conveyor can be used to hold the cans solidly in place as they ride on the conveyor's surface and are inverted. Once potential contaminants are removed, the container is reoriented in a position that facilitates filling, and sent on to the filler (hopefully on a conveyor that is covered to prevent later contamination). By placing a fine-mesh screen underneath the container flow path as it is inverted and reverted, contaminants that are washed free may be captured and examined.

3.6.2 Dealing with separation of FM from continuous roll-fed packaging films

Roll-fed packaging film stock produced for use as tray film or in heavier gauge form fill seal (FFS) applications can be liberated of foreign materials through a deionizer and air knife arrangement. Prevention of FM incidents can be aided in these applications by discussing the techniques used to wind the material at the vendor. Once the surface that was in the "up" position is identified (having an exposed area to the environment where dust or objects fall or settle upon at the time the roll was wound at the factory), practical precautions may be taken prior to use at the food processing facility.

If the film is homogenous in composition, it can be oriented such that the "up" surface is inverted as it is unwound (i.e., rotate the roll along its horizontal access 180°). If this is impractical due to film composition or printing, the unwinding and threading operation can be adjusted at the point of use to facilitate inversion. This could be adjusted via configuring various tension rollers in a manner to ensure that the surface is inverted at some point before being introduced to the product.

3.7 Monitoring physical separation processes

Obviously some of the physical separation methods outlined above could make for a difficult time weeding through the reject fractions. It is important, however, to take the time to document and analyze the reject fractions closely from the separation processes that can be monitored, such as screeners, air classifiers, strainers, and packaging rinsing. Both the FDA and USDA routinely ask about reject fraction analyses in their inspection activities, and processor compliance is a prerequisite to keeping the doors open for future business.

From a purely cost-containment perspective, the monitoring of reject fractions is a very worthwhile endeavor. Product that has been rejected is a point of financial loss for the corporation whose cost is not entirely recoverable in any situation. Many times material loss has to shift greatly before it is noticed by the accounting function during the process of a monthly or quarterly review, and gradual shifts over periods of time can result in financial bleeding for the company. Discovering that a shift in material waste has occurred long after the fact affords no opportunity to manage losses in a preventative manner. Putting systems in place to monitor the reject fraction allows the quality assurance professional to balance quality and product safety interests while helping to drive the business forward.

Placing a reject inspection system in place at the appropriate stage of the process requires knowledge of the process, previous historical data, and personnel to perform and document the inspections. The absence of all but personnel are usually able to be obtained from performing a HACCP-based risk assessment and data analysis to understand the typical mean time between the presence of failures (when contaminants are found) at the separation system. The operators who will perform the inspection of the rejects must be trained and given documented work instructions that detail the inspection procedure as well as what to do when contamination is found. Every inspection should be documented with positive confirmation that there was, or was not, FM finds. This document should be reviewed for each day of operation and reconciled with any product quarantines that were issued to ensure that each find was given the proper disposition (ship or hold).

Any contamination identified should be placed in a sample container and tagged with a brightly colored FM incident tag by the inspector at the time it is found. The tag should be filled out by the inspector to identify (1) the equipment where the FM was found, (2) date and time of the inspection, (3) name of the inspector, (4) name of the person in authority who was notified, (5) if there was a product quarantine ordered and the quantity of product quarantined if one was issued, and (6) whether there is a known assignable cause for the contaminant's presence (i.e., a dropped screwdriver). Over time the results can show a trend, and be used to highlight emerging issues or drive continuous improvement activities (procedures for trending results are presented in the acceptance sampling and statistical process control (SPC) chapter (Chapter 9) of this text).

3.8 Integrating separation processes into regulated HACCP plans

From a regulatory perspective, referencing any physical separation operation in a process as a critical control point within a federally mandated HACCP plan requires several criteria to be met. Typically a separation step is to be defined as a CCP only when, through its action, it prevents food items which contain foreign material contamination that could be injurious to public health from being distributed to consumers, or otherwise renders the physical hazard harmless. If a separation step is defined as a CCP in this regard, then a reasonable scientific justification that displays its effectiveness must be included in the plan.

In practice this means that the processor must show that the separation device effectively renders the FM (identified in the hazard analysis) harmless, either by culling it out of the product stream or by preventing the further processing of products that may contain the FM. Realize that this discussion is provided for FM that meets the definition for representing a hazard of public health significance detailed in Chapter 1. Typical evidence that can be used to successfully meet the effectiveness challenge in both USDA- and FDA-regulated environments are presented below:

1. Validation studies that are performed in situ and prove the equipment is effective in removing the contaminant in a repeatable manner. The results of these studies must be stringently documented and statistically valid to be accepted by the agencies.
2. Specific literature references from scientific studies published in reputable journals illustrating that the technology, as installed in a like-for-like manner in the facility in question, is an effective means of controlling FM.
3. Grandfather-clause-type "proof of validation through use" evidence collected over a long period of time while the separation process has been present in the facility. The write-up should include relevant facts such as (a) time-period equipment has operated continuously since installation, (b) the types of products that have been produced with this configuration, (c) data displaying that there is no history of consumer complaints for the type of FM identified in the hazard analysis when

the separation equipment is used, and (d) data displaying no FM-related reported deaths, injuries, or illnesses from use of the product produced while the separation strategy has been used.

Realize that even if the equipment suitability can be proven through the data presented, ongoing verification activities to ensure that the device is in a state of operational effectiveness, as designed, are required. This ongoing verification activity will need to be conducted and documented over periods of operation. This can be a challenge for some separation equipment, but in the case of the typical screener or strainer, the verification activity could be as simple as:

Ensure the correct screener/strainer is properly installed in the housing by checking the serial number and engraved mesh size.

The verification step could further detail to:

Measure the air gap between the housing and screener/strainer retention ring. Visually examine the screener/strainer for breaks, discontinuities of the screener/strainer material, or enlargement of the mesh size. Ensure operators are performing tailing inspections properly, and documentation conforms with the HACCP plan.

The specified limits for such verification activities could be:

1. Specified screen size and serial number shall be present and have no visible air gaps in the housing-to-screener interface
2. Zero voids, breaks, or visible enlargements of mesh size are allowed on the screen's surface.
3. Periodic documented audit of the inspection process to ensure it is performed according to the established procedure and frequency requirements.

3.9 Factors that can cause separation processes to fail

There are numerous reasons why separation strategies sometimes fail. The failure can be twofold—either the equipment will increasingly reject or damage good product and make it unsalable, or the FM is not rejected and is passed on through the system. Listed below are the top 12 root causes of failure in separation systems. The list was compiled through examining recalls, personal experience, and frank networking activities with numerous other processors throughout the food industry who have experienced failure. This list should be used by the reader to live vicariously through the mistakes of others, and to avoid feeling the pain of failure first hand. The top 12 reasons (not in any specific order) that separation strategies fail are as follows:

1. Failure of equipment through wear or catastrophic events
2. Improper redesign of material introduction or take-away systems (system out of balance)
3. Inconsistencies in utilities (air, water, electricity, or steam)
4. Unanticipated and unquantified changes in product characteristics (bulk density, moisture content, static charge, etc.)
5. Changes of suppliers
6. Increased throughput demands not compatible with the separator's capacity
7. Maintenance events in the area (planned and unplanned)
8. Untrained personnel introduced into the area (new hires, contractors, maintenance personnel, kaizen groups, etc.)
9. Unanticipated changes in environmental factors (temperature, humidity, or barometric pressure)
10. Leaks in the product flow path or separator itself usually continuously cleaned up by the operators in the area (good for sanitation, but bad for costs)
11. Lack of product quarantine procedures, defined chain of command for disregarding a defect, or other insufficient guidelines at the operator level
12. Reject material being reintroduced to the line due to:
    a. intentional practices to increase yield because there is disbelief that the equipment functions properly and that reject materials are still good
    b. not running rejects into inedible bins and having them unknowingly reworked onto the line after the separator is used

Works Cited and Select Resources

1. Alamzad H. 2004. *Vibratory Screeners: Spotting and Solving Premature Screen Breakage*. Powder and Bulk Engineering (3):67–71.

2. Farnell D, Thompson S. 2001. *Trommel Screens—Their Sizing and Design.* Montreal, QC, Canada: Farnell-Thompson Applied Technologies Inc. 10 p.

3. Huang CC. 1996. *Air Classifiers: How They Work and How to Select One*. Powder and Bulk Engineering (12):69–81

4. Ricklefs RD. 2000. *Selecting a Sifter for Scalping, Removing Fines, or Grading*. Powder and Bulk Engineering (12):19–31.

5. Thomas J. 1990. *An Introduction to Gravity Separators*. Powder and Bulk Engineering (12):36–38.

CHAPTER 4

Applications of Magnetic Separation to Prevent Foreign Material Contamination of Finished Food Products

4.1 Introduction

The contamination of a food product with metal can occur at any point in the supply chain, from farm to table. The points where metal contamination is reasonably likely to occur should be identified by the processor via conducting a hazard analysis in accordance with the principles of HACCP, as discussed earlier in the text. From an engineering standpoint processing steps within the supply chain that share the common trait of being subject to magnetic metal contamination can be dealt with in a similar manner. Many times the appropriate engineering solution is to address the relative risk of contamination by using modern magnetic separation techniques. To this day, magnetic separation is an underutilized application for removing magnetic metal contamination from food products.

Magnetic separation of contaminants from food products is a mature technology used at various stages throughout the food industry. Applications of magnetic separation in the food industry have been used for quite some time to sequester contamination from food products. The earliest iterations of the 20th century involved the use of weaker antiquated magnetic materials to identify gross contamination in grains and various other agricultural products (7). Use of magnetic separators on harvesting equipment, feed bunks, and cow magnets, as discussed in Chapter 2, are applications of magnetic separation that probably aid a majority of food processors.

The use of magnets for culling metal from food sources was broadly applied in Europe at the conclusion of World War II. Crop fields were heavily seeded with nearly a decade's worth of ordnance, shrapnel, and munitions from battles between the Axis and the Allies. Needless to say the prevalence of this magnetic burden made encountering metal contamination in produce, or in the stomachs of grazing animals, reasonably likely to occur. To reduce this burden the processors of the era started to suspend large magnets over the process lines to help pull this material out prior to further processing. Most recently advancements in the manufacture of magnetic materials has allowed for various types of magnetic separators to be developed for potential use in all areas of the industry.

Obviously the separation of contaminants from product streams can benefit the processor in any of four ways:

1 Prevent magnetic contaminants, that represent a hazard to public health, from reaching the consumer
2. Prevent products from being classified as adulterated by regulator agencies by removing excessive amounts of microscopic magnetic contaminants.
3. Use of powerful magnets can create a value added product that can be marketed based on its purity.
4. Prevent damage to costly equipment used in harvesting and processing.

Granted that magnetic separation is not as flashy a technology as automated x-ray inspection or continuous metal detection, but it has many advantages over these technologies. Magnetic separation can effectively eliminate a higher level of baseline metal contamination at a minimal cost to the producer (represented by waste or equipment) compared to detection methods that use reject mechanisms. Ideally, magnetic separators would be used in concert with metal detection and x-ray inspection equipment. From a strategic standpoint it makes a lot of sense to reduce the rejection burden on inspection units to ensure that truly contaminated product is removed from the stream. This is especially true when contrasting the methods by which the contaminants are removed from the process stream. Magnetic separators can continuously draw contamination from a process stream without having to reject large amounts of product. In contrast, detectors use mechanical reject mechanisms to cull out portions of the product stream that are potentially contaminated.

A relevant example of this concept can be found in the ingredients industry. If a 30 kg bag of dry enzyme (somewhat fluffy in texture) were to be rejected for ferrous contamination at the final step of the packaging line by a metal detection unit, then a magnetic separator might have proven more useful. Depending on the size of the contaminant and depth of the product burden, it could have easily been culled at an upstream process step by a magnetic separator. The cost advantage in this example would have to go to the magnetic separator . . . especially if there is no way to rework the finished product.

4.2 Principles of magnetism and materials utilized in construction of permanent magnets

Before installing magnetic separators to control metal contamination, it is important to understand how magnets work. The general principles that govern what types of material can or can't be used to manufacture a magnet, and what types of material can be separated from the product stream, are rooted in the science of quantum physics. Developing a basic conceptual understanding of how and why a permanent magnet works will afford enhanced comprehension of its limitations.

4.2.1 Magnetic magic demystified at the atomic level

Substances that are attracted to a magnetic field are called paramagnetic. To understand this principle of magnetism, one must start at the atomic level and build the concept to the macroscopic level. Essentially, any object of mass is composed of atoms of various elements. In theory each atom is composed of a myriad of subatomic particles that are identified and described by various models and can be represented with mathematics.

The atom, as represented in Bohr's model, describes the nucleus containing protons (with positive charge) and neutrons (with no net charge). Electrons are negatively charged particles, and their negative charges help to bring stability to the net positively charged nucleus. Electrons exist in a "cloud" outside of the nucleus and, according to the Heisenberg uncertainty principle, their exact location cannot be pinpointed. By using quantum mechanics, however, we can describe the area that an electron occupies with various constants (n, ℓ, m_ℓ) and the paths, or orbitals, that they could travel within in a given situation are dependent on the energy state of the atom. The electrons can exist in these various orbital states—each characterized by a variance in distance traveled from and an overall path taken around the nucleus (6).

Each electron within an atom possesses a specific magnetic spin constant (quantum expression of m_s) that can be in any of two mutually exclusive states, either $+\frac{1}{2}$ ($\uparrow$) or $-\frac{1}{2}$ ($\downarrow$). This phenomenon has also been described by the term "polarity," and we can illustrate this concept by referencing Figure 4.1. There are specific rules that govern how many electrons of each type of spin can be present within an orbital level. These rules also dictate how an electron will be promoted (in the presence of excess energy) or demoted to a different orbital when the atom is subjected to external forces. If more than one electron is to coexist within an orbital, they must possess opposite spins from one another. When this condition exists the electrons are considered paired (expressed in symbolism by $\uparrow\downarrow$) and their magnetic spins constants are offsetting. Substances composed of atoms that possess electrons in the paired state are generally

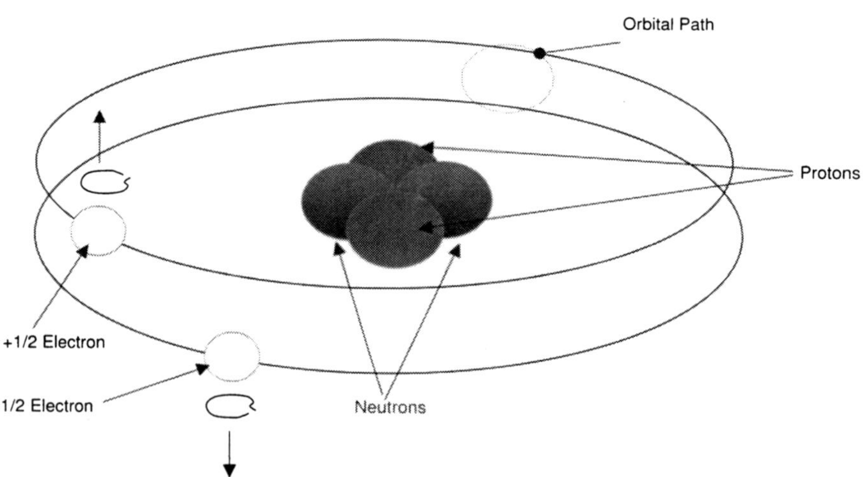

Figure 4.1. Components of a typical atom, and an illustration of electron pairs that can co-exist in an orbit due to their opposite spin constants.

diamagnetic, and therefore do not react to magnets (6). An orbit may also be occupied by a single unpaired electron that can be of either spin configuration (↓ or ↑) as long as another electron of the same spin is not promoted to that level.

Generally speaking the substances composed of atoms with unpaired electrons (Figure 4.2) are paramagnetic in nature and tend to interact with external magnetic fields. Essentially, paramagnetic atoms behave as magnets themselves and tend to have North and South poles (much like a bar magnet). When paramagnetic atoms are exposed to external magnetic fields, the unpaired electrons assume a common spin configuration that is parallel to the field applied. This phenomenon is sometimes referred to as an induced magnetic moment. After the magnetic field is withdrawn, the unpaired electrons of most elements will revert back to their nonmagnetized spin configurations. Elements that are classified as ferromagnetic, however, will actually retain the spin configuration (either ↑, or ↓) induced even after the field is removed. The electron spins within the orbitals of iron (Fe), a common highly paramagnetic metallic element, are models of this characteristic behavior. Iron has four unpaired electrons available for attraction to a magnetic pole, and it is easy to see why iron atoms are easily influenced by magnetic fields.

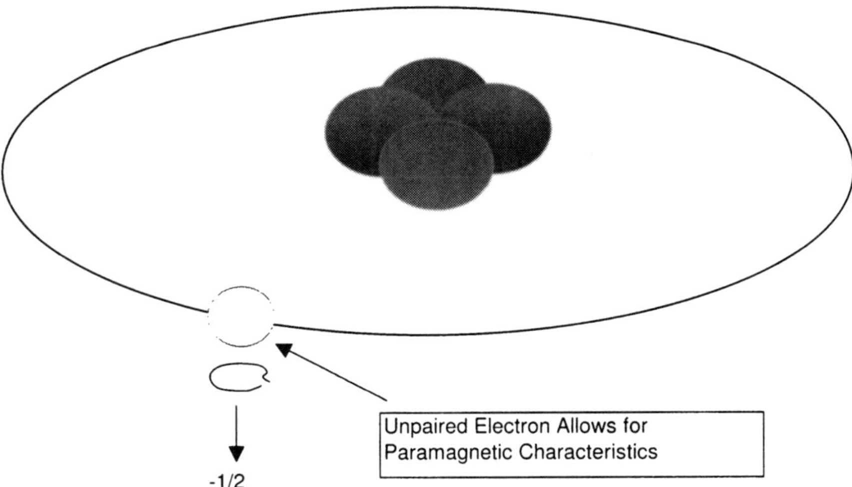

Figure 4.2. Illustration of an atom that possesses an unpaired electron whose spin can be induced by a magnetic field.

4.2.2 Magnetism at the larger scale

Applying the concepts of magnetism learned at the atomic level (defined above) to everyday materials composed of differing atomic constituents allows one to grasp magnetic materials in application. Let's examine a 500 g ingot of pure iron to illustrate this concept. By using the average atomic mass unit for iron (obtained from the periodic table of the elements), one can calculate the number of iron atoms present in the ingot. The calculation indicates that there are 5.39E+24 individual iron atoms in this ingot arranged in the form of a complex crystal structure (a common preferred structure for all metals). Some of the iron atoms within the lattice structure are going to have multiple unpaired electrons that will be eagerly available for interaction with magnetic fields. The magnetic spin of the unpaired electrons are randomly distributed between ↑ and ↓ throughout this ingot, but when we expose this iron ingot to a strong magnetic field large groups of the unpaired electrons within the lattices will be influenced to align

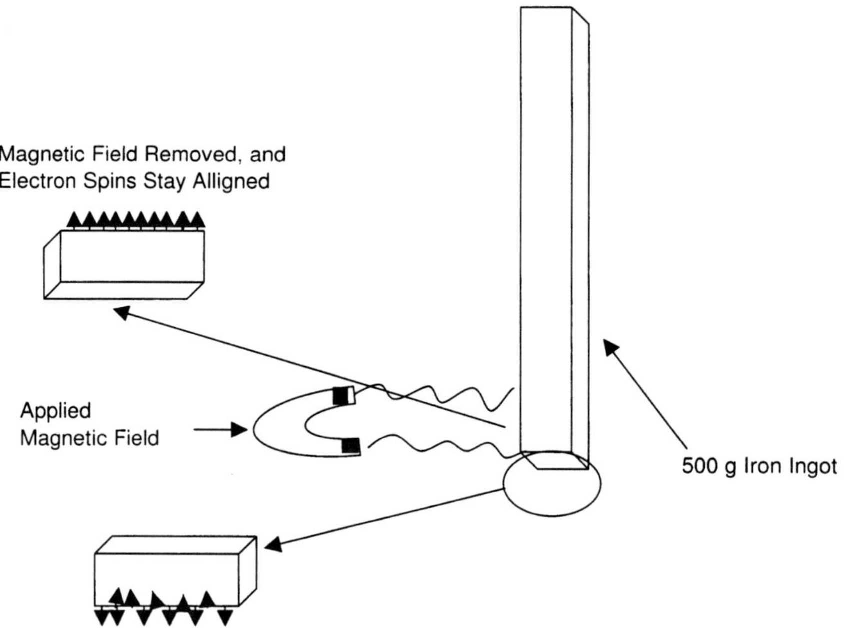

Figure 4.3. Alignment of unpaired iron electrons subjected to a magnetic field.

with the field. The magnetic field can now be removed, and the unpaired electrons of the iron will retain their new spin characteristics thanks to their ferromagnetic nature, and will now exude a stronger magnetic field of their own (Figure 4.3).

This newly formed iron bar magnet can then be used to influence other paramagnetic compounds in the same manner. The magnetic properties of the iron bar magnet will remain until it is demagnetized (i.e., the electron spins are shuffled back to random). This can be done by heating the iron up to the point where the crystalline structure is lost (i.e., transition/melting), or by reacting the iron with other chemicals that bond with the electrons.

Compounds used in modern material science are composed of various elements whose atoms are arranged in various geometric structures that may or may not contain free electrons. The attraction that a material will display to a magnetic field is based upon this structure and other factors, like the contaminant's size. Note that on most occasions the contaminants that a typical magnetic separator are likely to sequester are iron, rust, magnetic minerals, and certain grades of stainless steel. In some cases rocks and stones that are rich in magnetic minerals can be attracted to a magnetic separator and prevent costly equipment damage.

4.3 Why stainless steel can sometimes be culled by magnetic separation

The underlying reasons why stainless steel can sometimes be culled from food products by magnetic separation are poorly understood by most people in the food processing industry. Many believe that all stainless steel is unaffected by the application of a magnetic field and therefore not subject to magnetic separation. This statement is false. In order to debunk the paradigm, it is necessary to thoroughly discuss the metallurgy of stainless steel and its production processes.

4.3.1 What is stainless steel?

Stainless steel is manufactured when chromium and other selected elements are added to a molten mixture of iron. The types of compounds normally added to stainless steel and the benefits of each are described in Table 4.1. By being able to create steel of varying characteristics, modern metallurgy and material science has advanced at a rapid rate. Various types and series of stainless steel available are presented in Table 4.2. Note that most of the desirable grades for use in the food industry are in the 300 series of steel. Clearly many of the characteristics, especially those that aid in cleaning soils from surfaces and preventing corrosion and leaching of metals, are beneficial to the food industry.

Evaluating stainless steel via its series number and characteristics is also helpful when attempting to ascertain the suitability of a material for a specific application. A deeper understanding of the metallurgy beyond the series number nomenclature, however, must be understood by those who are seeking to grasp the concept of why certain types of stainless steel are attracted by a magnet, and others are not. Essentially all varieties of stainless steel can be categorized as either belonging to or being a hybrid of three basic types: ferritic, martensitic, or austenitic. Each type is dif-

Table 4.1 Compounds typically utilized in the manufacture of stainless steel, and the behaviors each invokes.

Alloy Elements	*Functional Benefit of Use*
Aluminum	Forms strong ferrite, improves scaling resistance, and lowers ability of the steel to harden
Carbon	Decreases resistance to corrosion, but adds to strength
Chromium	Chromium oxide surface film increases corrosion resistance and heat-scaling resistance
Columbium	Promotes formation of ferrite, interacts with use of carbon to decrease corrosion tendency
Copper	Increases corrosion resistance in certain situations, provides age-hardening effects, decreases susceptibility to stress corrosion cracking
Manganese	Adds to austenite stability at room temperatures; is cheap partial replacement of nickel and uses >2 percent; can affect strength, ductility, and overall wear resistance
Molybdenum	Increases corrosion resistance and strength at higher temperatures, counteracts pitting
Nickel	Stabilizes austenitic structure, increases ductility, increases strength at elevated temperatures, and increases resistance to corrosion
Silicon	Slightly increases hardness and tensile strength, increases scaling resistance, resists carburizing at elevated temperatures
Titanium	Interacts with use of carbon to decrease corrosion, promotes ferrite formation

Table 4.2. Comparison of commercially available stainless steels: composition and properties.

Series Number	*Stainless Type*	*Characteristics*	*% Nickel*	*% Chromium*	*% Carbon*
410S	Ferritic	Cheap, better strength and corrosion resistance than iron	N/A	12	0.015
430	Ferritic	Corrosion resistant, good strength	N/A	16.5	0.05
444	Ferritic	Oxidation resistant, corrosion resistant, and excellent stress profile	N/A	17.5	0.015
410	Martensitic	Can be readily hardened, good for cutlery	N/A	11.5	0.14
420	Martensitic	Better hardening than 410, excellent for knives and cutlery	N/A	12.5	0.38
304	Austenitic	Corrosion resistant, oxidation resistant, and readily shaped	8	18	0.06
316	Austenitic	Resists corrosion, pitting, and oxidation, and can be readily shaped	10.5	16.5	0.05

Chemical composition taken from publicly advertised mean values for each product manufactured by AK Steel Corporation. Middletown. Ohio.

ferentiated by possessing a characteristic spatial distribution of the atoms that they are composed of. This spatial distribution is referred to as the microstructure of the metal. The nature of how the atoms are arranged in each microstructure will allow us to determine which stainless steels will be a candidate for magnetic separation.

Generally iron crystals and their stainless steel alloys have two general orientations that their microstructures can assume when in solid form. The atoms of the metals can be organized in either a face-centered cubic structure (FCC), or a body-centered cubic structure (BCC). We will detail what these are soon, but for a quick reference refer to Figure 4.4. FCC structures are an unnatural structure for iron atoms in steel to assume, and it normally requires a large amount of external energy, in the form of heat or pressure, to maintain iron atoms in the FCC microstructure. Essentially maintaining an FCC microstructure in an iron steel is the human equivalent to coercing 14 people to simultaneously take a road trip in a compact car . . . it can be done, but it requires a large external motivational influence. Typically BCC structures are more natural and are the preferred structure for iron steel to maintain. In fact the amount of external energy necessary to cause an iron steel (ferritic) to transition from a BCC structure to an FCC structure is not normally achieved until it is heated to 912°C (8).

The microstructure of austenitic stainless steel is unique, however, because it is FCC at room temperature. The majority of stainless steel used in the food industry is of an austenitic variety. Austenitic stainless steels contain chromium, increased levels of nickel, and reduced levels of carbon—as cast, they have extremely low-magnetic permeability. Austenitic stainless steel exhibits excellent resistance to oxidation and subsequent corrosion in high-temperature or acidic environments, and can be easily polished because it is highly ductile. The soft nature of austenitic stainless steel makes it ideal for being shaped, rolled, or pressed into differing conformations (think of tanks, vessels, and ductwork) (8). The reason for this extended workability and lack of magnetic permeability is found when its unique microstructure is examined.

Austenitic stainless steel varieties possess face-centered cubic structures at room temperatures. This means that a three-dimensional cube exists with an iron atom present in each corner of the cube as well as an iron atom being present in the center of each face of the cube. There are eight corners in a cube, so there would be eight iron atoms present at this point.

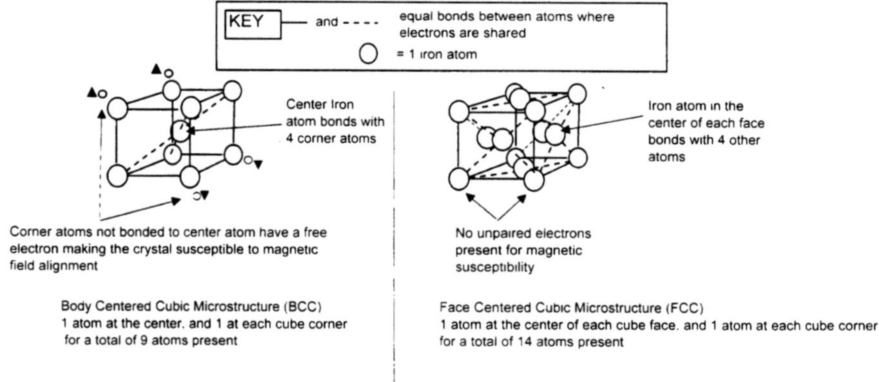

Figure 4.4. Body-centered cubic (BCC) structure compared to face-centered cubic (FCC) structure of atoms in iron crystal.

A cube by definition has six faces, so an additional six iron atoms could be present in this configuration, bringing it to a total of 14 iron atoms per cube.

As previously mentioned, this unique structure usually exists only at very high temperatures; however, austenitic steels contain measurable amounts of nickel which acts to stabilize the FCC structure at lower temperatures. The steel possesses an extremely weak influence to a magnet in this structure because, when the iron atoms are bound in this manner, there are very few unpaired electrons present to align with the field. The electrons are no longer in the unpaired state because they are involved in making bonds with other iron, nickel, and chromium atoms to a lesser extent. In order to get this structure to remain stable at room temperature, a measurable percentage of nickel must be dispersed throughout the steel. Nickel atoms will be randomly substituted into the structure anywhere an iron atom would normally be. As a general rule the higher the nickel content contained within a steel alloy, the lower reaction to magnetic fields it will possess, and its ability to conduct electricity is also reduced. This fact is very relevant to metal detection scenarios.

Martenistic stainless steels, in contrast, are influenced by magnetic fields and can be candidates for magnetic separation. Martinsitic stainless steels are very hard because they contain measurable amounts of carbon atoms that are internalized in their crystal structures. The carbon is internalized through its addition at high temperatures followed by an extremely rapid cooling to ensure that it remains entrapped in the BCC crystal structure (6). The BCC structure, even with the bonded carbon, still possesses free electrons that can align with a magnetic field. Martenistic stainless steels are usually used in the manufacture of knives for chopping, slicing, and dicing food products due to their extreme resilience to breakage, deformation, and wear tolerance (dulling).

When work hardening occurs at its surfaces, as in the manufacture of a typical fastener, austenitic steels can become susceptible to being magnetized. This occurs because the hardening, or working of the metal, can dislocate the crystal lattices into subformations of their smaller elemental constituents. The lattices also become more randomized in direction to form a different grain structure, which can once again yield unpaired electrons. This new grain structure will have martenistic BCC arrangements, and be susceptible to magnetic fields. An example of how this can happen can be found in a typical screw conveyor setup using 304 stainless components. The screw can rub against the metal housing, resulting in metal-to-metal contact, and generate either fine flakes or spiral wire type (swarf) stainless metal contamination. This type of contaminant, due to the nature of its generation, can be a candidate for magnetic separation in product matrices that don't present a large burden of travel. Use of powerful rare-earth magnets are especially helpful in this situation.

4.4 Common terms used to classify the properties of magnets

Obviously, magnets are conceptually difficult to present simplistically. This is especially true when trying to compare the strength of permanent magnets of differing composition or origin of manufacture. The concept of understanding the strength value of a magnet expressed in numeric terms is difficult and sometimes subject to skepticism, due to the way it is measured by the producer. Understanding the relative strength of a magnetic field can be confusing and difficult to interpret because the proper calculation of magnetic force involves a spatial component. Once a spatial component is introduced, the use of vector mathematics is necessary to model the result. This is essential because the pull felt by a ferromagnetic material within a field is dependant on its relative position and distance from the magnet. Understanding an expression of temperature, for instance, is much simpler because it is not a vector. Temperature, like most other expressions that people are used to dealing with, is two-dimensional in nature and can be expressed linearly in units of degrees. We can all relate to the fact that −4°C is a cold environment, and 50°C is an extremely warm environment. Unfortunately, when people hear that the "energy product of our magnets is in excess of 400 gauss oersteds," this is a pretty obscure concept to most.

So how does one tell how strong the magnetic field of a permanent magnet is at a given distance from the magnet? Elementary physics states that the magnetic field strength of a magnet, being a form of electromagnetic radiation, is approximately proportional to the inverse cube of the distance from the surface. Therefore, if you double the distance from the magnet, the magnetic field strength will be roughly reduced by a factor of eight.

Commercially available magnets are classified by their coercive force Hc (expressed in oersteds), residual induction Br (expressed in gauss), and their maximum value of energy product (BH)max (expressed in mega gauss oersteds, or simply MGOs) (10). Manufacturers of magnets can typically deliver tolerances of +/− 8 percent Hc and +/− 10 percent (BH)max (10), so advertised values need to be somewhat scrutinized. Comparison of permanent magnet compositions and their relative strength are characterized in Table 4.3. Figure 4.5 illustrates graphically the ranges in magnet strength present in the various types of magnets commercially available.

4.5 Composition and historical uses of permanent magnets in the food industry

The modern history of permanent magnets started about 1940 with the introduction of Alnico compositions. Before this, the use of permanent magnets was limited to those few applications whose very function depended on their permanent magnetic properties (instrumentation, compasses, and magnetos). During this period the use of magnets started to become widespread in devices such as motors, generators, and audio speakers. The strength of the early magnetic separators used in the food industry was limited because of the materials that were used for their construction.

Table 4.3 Comparison of commercially available magnetic materials by their relative strength measurements.

*Type of Magnet**	*Typical Coercive Force (Hc) (oersteds)**	*Typical Max Energy Product (BH) max (MGOe)**
Alnico 1	470	1.4
Alnico 6	780	3.9
Alnico 9	10600	9
Ceramic 1	1860	1.05
Ceramic 5	2400	3.4
Ceramic 8	2950	3.5
Rare Earth (Samarium Cobalt)	8500	18
Rare Earth (Samarium/Cobalt with Praseodymium)	9000	22
Rare Earth (Complexed with Neodymium and Dysprosium)	12000	38

*Types and values obtained from guidelines published in *MMPA Standard No. 0100–00 Standard Specifications for Permanent Magnet Materials.*

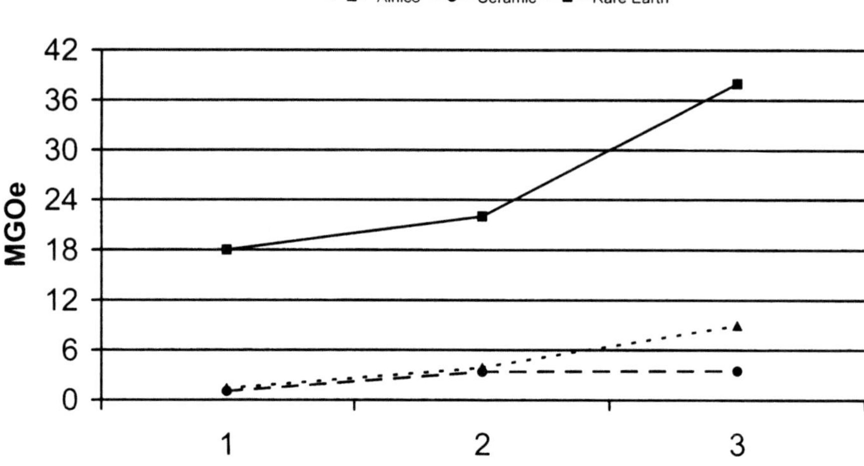

Figure 4.5. Strength comparisons of various families of magnetic materials.

For this reason some processors would incorporate bulky and expensive electromagnets into their process lines to address FM. Ferrite permanent magnets were used until Alnico magnets became available at costs that were affordable for use by food processors in the 1970s.

In the 1980s rare-earth magnetic separators became available at costs that were beginning to be compatible with the budgets of food processors, thanks to the increased demand for their use in the technology sector (12). The discovery of neodymium-iron-boron magnets was announced in the early 1980s by Sumitomo Special Metals Company and General Motors. These rare-earth magnets were based on an iron structure with a neodymium-boron compound dispersed within it. This new magnetic material is stronger than previous existing types of rare-earth magnets, and they are less expensive because they are made primarily from iron—which is much cheaper than sourcing cobalt and samarium of sufficient purity to make magnets.

Today rare-earth magnets are still more costly than Alnico and ceramics, but this cost discrepancy has steadily decreased. The proliferation of powerful rare-earth magnetic separators will surely continue as prices decrease and more unique and efficient designs become available for use.

4.6 Types of magnetic separators available for use in the food industry

There are basically four categories of magnetic separators available for use in the food industry. Each type can be composed of any kind of magnetic material from Alnico to rare-earth, and they are differentiated by their construction and potential application. The four categories of permanent magnets used in the food industry are:

1. Plate magnets
2. Pipeline magnets
3. Rod magnets
4. Drum magnets

Each type will be discussed in the sections outlined below. Understanding the potential application areas for each type is essential for using magnetic separators to prevent magnetic FM in food products.

4.6.1 Plate magnetic separators

A plate separator is simply a fixture (typically a box, metal disk, or plate) that contains permanent magnet(s) that are mounted to its surface(s). A plate magnet can be installed above the flow of the product or in contact with the product stream. A typical installation for a plate magnet is to be suspended above a conveyor to catch contaminants as the product passes underneath. The plate magnet will need to be sized appropriately for the width and speed of the conveyor to ensure maximum exposure. In addition, the plate will need to be placed as close to the surface of the product as possible to ensure a maximum field exposure (due to the loss over distance). Figures 4.6 and 4.7 display typical suspended plate magnet configurations used to cull magnetic FM.

The depth of the product needs to be managed carefully as well. Ideally the product would be very shallow in its distribution at this point because the magnet will have to attract paramagnetic particles, potentially from the surface of the conveyor, and effectively manage to pull them through the product bulk to its surface. Effectively managing product depth can be accomplished by numerous mechanical methods. A simple plow affixed above the conveyor at the desired product height is an effective tool as long as it doesn't create a bottleneck for the manufacturing process. Other methods that are effective are using a wide, slow-moving conveyor for the

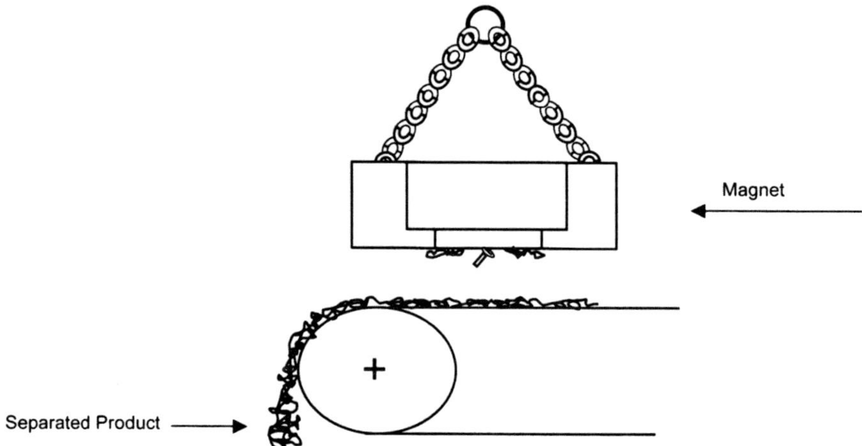

Figure 4.6. A possible configuration for mounting a suspended magnet over a product conveyor.

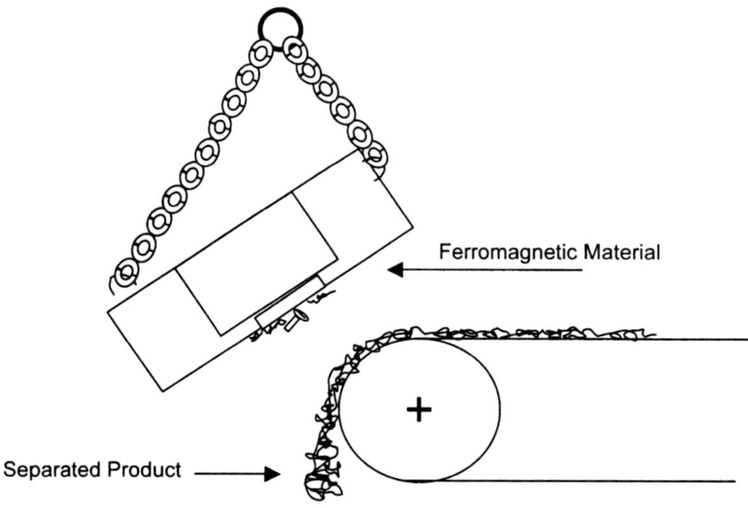

Figure 4.7. Another possible configuration for mounting a suspended magnet over a product conveyor.

magnetic separation, and fanning the product onto it from a narrower faster-running conveyor.

Plate magnets can also be placed in the product flow, potentially in contact with the product as well. A typical example of this type is a hump magnet. In this arrangement, material is allowed to flow through an area of pipe or ductwork; at a bend area in the flow path a permanent magnet awaits to draw contaminants from the product. Placing the magnet immediately after this bend in the transfer path is advantageous because a change in direction usually represents a decrease in velocity of the material, affording a greater contact period with the magnet. The hump also helps to direct the material stream closer to the magnet so its effectiveness can be optimized. The magnet is usually placed on a hinged metal plate with a gasket or seal around its circumference to prevent leakage (Figure 4.8). Using a hinged setup affords easy access for inspection and cleaning.

Another typical use is to place plate magnets at the bottom of discharge chutes. By placing the magnets in the discharge location, there is a good chance that magnetic contaminants accumulated within the product will be captured and retained on the plate surface as they flow over it. Ideally the plate angle at the chute discharge should not be greater than 45° to facilitate maximum exposure of the plate surface to the product. Again, the plate will need to be sized for the flow rate of the operation. When mounted on a hinge this design also allows for easy access and visual monitoring by the individuals tasked with such activity.

A wedge formed out of plate magnets and placed in the product flow of a free-falling duct, or pneumatic conveying operation, is another common application for separation of paramagnetic materials. Essentially product is allowed to flow over the wedge via gravity or pneumatic action. As the material contacts the wedge surfaces, magnetic contaminants are captured

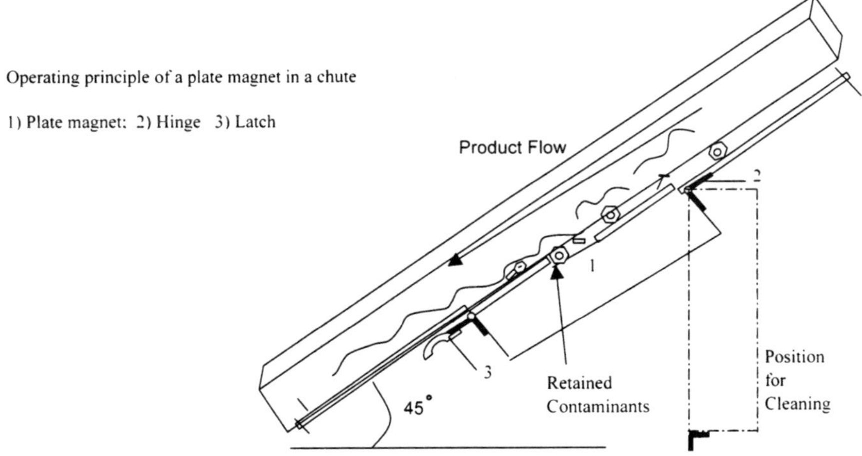

Figure 4.8. Operating principle of a plate magnet in a chute.

and retained on the surface. The wedge can be removed from the process flow for inspection and cleaning activities via a hinged access panel and a sliding arm. Cleaning can be facilitated by placing a thin removable 316 stainless steel or aluminum "hat" on the wedge. The design of this system must ensure that the product will flow freely over the wedge, and changes in flow will not bias the material to flow in a manner that could potentially bypass the magnetic surface.

4.6.2 Rod magnets and other cylindrical magnets

Magnetic materials manufactured in a cylindrical form have many applications in the food industry. The simplest form of this type of magnet is the single magnetic tube. This design can be used like a wedge plate magnet or as a sampling device to gauge incoming material quality (insert clean rod in product, stir for prescribed time, and examine for contaminants) of various ingredients and packaging materials, such as loads of plastic resin. When used in the flow path of a product, a single magnetic tube can have limited effectiveness if sized inappropriately or not placed correctly to maintain maximum contact with the product stream.

Product exposure for rod-style magnets is gained when multiple rods are arranged in a grate configuration. Liquids or solids can flow through the grate, and paramagnetic contaminants will be captured on the surface of the rods. The magnetic material used to manufacture the rods, as well as the distance between each rod, will determine the level of contamination that can be removed by the grate. Magnetic grates must be cleaned periodically to ensure they remain effective. This used to be a tedious and time-consuming process for the folks who were responsible for monitoring the magnet. Contemporary versions of magnetic grate separators, however, are set up much like a drawer to ease in the cleaning operation. The actual magnets are housed inside a thin housing of diamagnetic material, many times made of stainless steel, that is in contact with the product. The whole grate is allowed to slide out of the product stream for visual inspection. If contaminants are observed and need to be removed, the magnets of the grate can be slid out from their housings and the paramagnetic contaminants will fall free into a collection area. Grate magnet setups are commonly used in areas where product ingredients are metered into the process. A good example of where a grate magnet is a good first line of defense is at a station where a person is cutting open 50-lb bags of flour and dumping them into a weigh hopper. Bulk liquids can also be poured through a grate separator to remove larger contaminants.

For maximum contamination removal with rod-style magnets, a series of grate magnets can be configured into a grid arrangement and placed in the product stream. A paramagnetic contaminant is faced with a torturous path to bypass the series of multiple grids. This type of arrangement works well with dry ingredients when placed within the ductwork of a process, or at the discharge of bulk bins/silos. The sizing of the grid, and the construction of the ductwork that leads to it, are extremely important to ensure that material flow is not compromised. If the material that is to be passed through a magnetic grid is notorious for clumping or bridging, a vibratory mechanism or a series of air injection nozzles at the separator may be necessary to ensure plug-ups do not occur.

4.6.3 Liquid traps

Rod and plate magnets can be utilized to cull contaminants from products that are pumped through piping (liquids, slurries, soups, salsas, etc.). The traps come in various designs, and they can be composed of Alnico or rare-earth components. Traps with multiple fingers are outstanding for removing fine contaminants from liquid lines that are devoid of particulates. Their effectiveness is dependant upon the material flow rates, viscosity, and spatial distance between the fingers themselves.

Installations incorporating a three-way valve splitting flow to two liquid traps allow ease of cleaning while providing uninterrupted removal of the contaminants. The traps can be checked safely while the process continues to run. Some newer models of traps offer dual functionality—by incorporating a single powerful rod magnet positioned inside of a filter screen, further physical separation of FM contaminants from products will result (3). The beauty of this system is that the operator can perform two verification checks at once by disassembling the same equipment.

4.6.4 Magnetic pulleys

Magnetic drums and magnetic pulleys are two types of separators that can make an impact in an operation seeking to decrease contamination risk and increase the level of product purity. As with other separation techniques the magnetic source may originate from a variety of permanent magnetic materials, or from an electromagnet (2). Magnetic pulleys are an effective control mechanism that are simple to install and maintain for an operation that already uses conveyors to move materials to or through process operations. Magnetic pulleys are also an excellent second line of defense against paramagnetic metal contamination when used either upstream or downstream of plate magnet designs.

Essentially, a magnetic pulley is simply a cylinder that is matched to the width of the conveyor belting and has a central shaft for support. Magnets are installed on the inside of the cylinder in a manner that ensures the cylinder exhibits a magnetic field strength with limited variation around its perimeter. Pulleys, which are normally mounted on the discharge end of a conveyor belt, can be ordered in different sizes to fit the specified application. When no contamination is present they behave just like a normal idle, or drive, pulley would in providing support and tension to the belt material of the conveyor (2). The difference is that when paramagnetic materials are present within the product flow on a conveyor belt, they are attracted to the surface of the magnet and should follow the belt path down and around the pulley against the force of gravity to be discharged into a collection bin. Figure 4.9 illustrates a typical magnetic pulley system used to counteract FM.

This is an ideal case provided there aren't constraints in field strength or depth of material present on the belt. The other diamagnetic product present is able to fall freely from the conveyor belt in accordance with the laws of gravity and is usually discharged onto another belt, into another piece of processing equipment, or into a hopper. Pulleys are self-cleaning by design and require limited attention by maintenance personnel. Real-

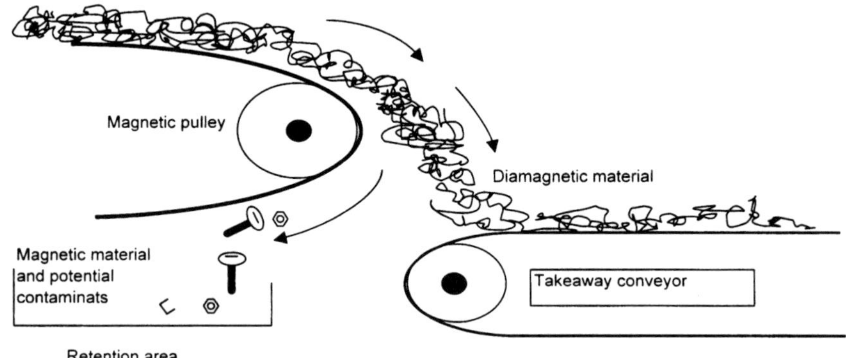

Figure 4.9. A common magnetic pulley design to remove tramp metal and other magnetic foreign materials.

ize, however, that the effectiveness of a magnetic pulley is constrained by several factors. The product depth and relative bulk must not impede the travel of materials to be held steadfast on the conveyor surface. The conveyor speed that the pulley was designed for is a third factor. Unlike a plate magnet, a pulley has a much smaller effective working surface area, so speed sensitivity is magnified in a pulley setup. Work with the pulley manufacturer closely, and bring as much product and line speed data as possible for their review prior to installing a system.

When selecting a pulley there are two other important characteristics to consider that are not related to its collection characteristics. Unnecessary wear and further product contamination potential can be avoided throughout the service life of the pulley by identifying that the shaft can accept a high-quality serviceable bearing. Additionally the external sleeve and surfaces of the pulley must be suitable for the day-to-day operational and cleaning environment where it will be placed into service. By following through on the above requirements, a long worry-free service record can be expected out of this separator design.

4.6.5 Magnetic drum separators

Magnetic drum separators are another great application for winnowing paramagnetic contaminants from dry or wet continuous product streams. Drum separators can be used in installations where plate magnets and pulley applications are not feasible or practical (1). Magnetic drums incorporate either a permanent magnet or electromagnet fixture retained in a fixed position around a rotating drum (9). The product can be fed to the drum surface by either a gravity feeder, a horizontal conveyance, and/or a liquid media flow onto or just below the surface of the drum. As the drum surface rotates across the magnet, the susceptible paramagnetic material is retained until it moves past the effective working area of the magnet. At this point the magnetic material falls from the drum surface and is discharged into a reject bin for later examination. Figures 4.10 and 4.11 illustrate common feed configurations for magnetic drum separators. The gravity feed setup is preferred for realizing more complete separation of contaminants, due to the fact that gravity is aiding the travel of the particle to the surface of the drum versus acting against it (as is the case in a horizontal feeding setup). Realize that drum separators have the same constraints as a magnetic pulley in terms of flow rates and interaction with product bulk. Maintaining a continuous cascading flow to the separator versus a

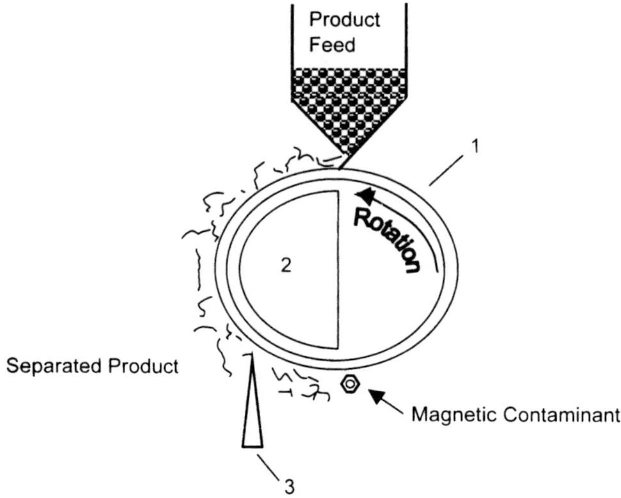

Figure 4.10. Illustration of an overfeed configuration for a magnetic drum separator.

pattern of discharge slugs of material is critical to maintaining effective separation.

There are a few traits that must be considered before using magnetic drum separators as a separation process. Drum separators are inherently more costly than plate magnets or pulleys due to their design. Drum separators require a larger physical footprint, and many times a redesign of the product flow system is necessary prior to installation in an existing process (9). Unlike a plate magnet, drum separators are not passive in nature and the drive mechanism requires a reliable power supply and effective preventative maintenance to retain its reliability. For ensuring the safety of personnel in the area of the separator and to keep product from being liberated to the environment, the separators are usually guarded by external housings.

As with any piece of shielded equipment, a thorough review of the sanitary design and methods for cleaning the separator must be developed prior to selection and subsequent installation. For microsensitive products, especially wet products of sufficient water activity to support pathogens, it is prudent to perform a microbiological monitoring study upon installa-

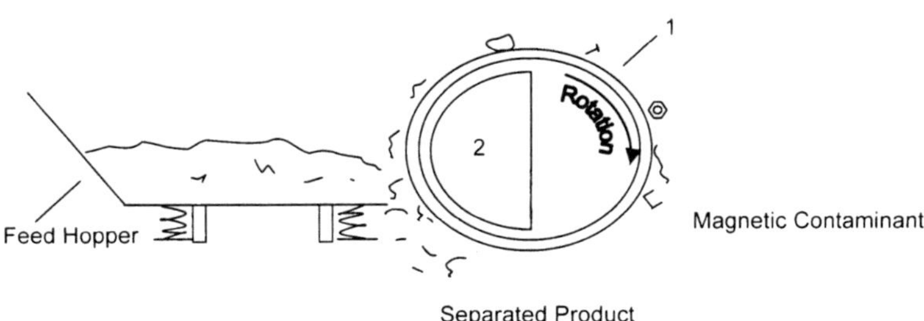

Figure 4.11. Illustration of an underfeed configuration for a magnetic drum separator.

tion of the equipment. The findings of the microbiological monitoring process may require that periodic sanitation activities are conducted at certain intervals of continuous operation. By performing such activities in a proactive manner, the processor can ensure that the drum separation operation doesn't become a point of biological contamination of the product.

4.7 Integrating magnetic separators into existing HACCP and QA systems

Purchase and installation of a magnetic separator is not an end-all to preventing paramagnetic contamination of food products. In order to maximize the efficacy of magnetic separation techniques, they must be integrated into a comprehensive food safety and quality assurance program. Hopefully the separator that is chosen, and its location of placement, will have been derived from the results of a scientific hazard analysis. Many times engineers responsible for implementing projects in manufacturing facilities will observe that an overriding corporate policy requires the installation of magnets on all process lines. To comply with the policy the engineer may purchase a plate magnet and consider the requirement met for the new operation, but the shrewd food safety and quality assurance professional will be involved in new process designs or retrofits upfront to ensure that optimal planning is done in accordance with the necessary risk mitigation.

Once the location and type of magnet are chosen and installation is complete, a flow chart of the process that clearly defines the location and type of separators must be developed and maintained. Obviously it is a requirement of HACCP to flowchart the process, but going beyond the square box descriptor that says "magnet" and detailing what type is there is a great upgrade. Using a graphic representation of the layers of separation technology within a process provides a much better tool for training employees about control points in the food safety or food quality systems. Displaying controls in this manner will also show those third parties who visit the facility a more disciplined and detailed presence of the facility's foreign material control program.

By possessing a drawing that identifies the type of separator present in each location, the processor has the added benefit of ensuring that change control is implemented. To better illustrate this concept, imagine a situation where magnetic separators are merely referenced as "magnets" in the process flow diagrams. Unknowing maintenance personnel and engineers may install the wrong type of "magnet" if a process retrofit or replacement is necessary. As with most system breakdowns it isn't the intent of these individuals to circumnavigate the systems, it is the clear lack of system definition and ownership that creates a negative outcome.

Each magnetic separator should be serialized in a unique manner to provide traceability. By giving each magnet a permanent and visible identification tag, the overall opportunity for using the wrong magnet in an area that it wasn't intended for can be reduced substantially. The processor can also more readily implement calibration programs and change control programs, and develop better-defined procedures dealing with the magnetic separator. Additionally, when the serial numbers are tied into a preventative maintenance system (PM), they are more readily audited to ensure the PMs are recorded and completed according to schedule.

4.7.1 Preventative maintenance and calibration of magnetic separation equipment

Typical PM for a magnet is to periodically inspect it to ensure that it has not been damaged, physically disturbed, weathered, or moved from the point of installation without approval. This activity should be done on a periodic basis, and be documented as an official record. Depending on the activity of the processing facility, the frequency for this inspection can vary from daily to semiannually. Additionally the inspection of magnetic separators should occur after any event which may have caused damage or undue movement of the separator.

Calibration of the magnet should be performed according to the recommendations of the manufacturer. Typically a gauss meter is used to check permanent magnets, so an individual trained in proper use of the equipment must be accountable for performing this activity. Magnetic drift is rare unless there has been a severe physical disturbance or exposure to excess heat. Common practice in the food industry usually requires only that the calibration of a magnetic separator be performed on a frequency of no less than an annual inspection.

4.7.2 In-process checks for contamination being present on the separator

The frequency with which a magnetic separator needs to be checked must be determined by the processor, and requires knowledge of the contamination potential and throughput rates. At no time should a non-self-cleaning magnetic separator be left unchecked for extended periods. This is important because material collecting on the separator will reduce the efficacy of the separator (11). Additionally the separator itself can become a point of metal contamination potential. This type of contamination occurs when the separator is saturated with paramagnetic materials and begins to lose retention as products, or the environment, disturb the contaminants retained. Tracking down the cause of periodic metal contamination can be maddening, especially when the root cause is a separator that has been "lost" to cleaning and has been collecting contaminants for months.

4.7.3 Written procedures for inspection of magnetic separators and documentation of results against clear standards

All of the effort expended in conducting a thorough hazard analysis and defining the use of magnets in the manner illustrated above would be nullified by allowing untrained personnel to inspect the separator and define the contamination present, by their gut feel, as being significant or not. Clear and well-defined work instructions for the periodic checking and reinstallation of magnetic separators must be developed prior to the beginning of production.

Work instructions should be included within the scope of a document control program so that nonapproved revisions may not be permitted. The work instructions should detail how to remove the separator from the process, and whether or not product flow should be stopped or diverted during this time frame. They should also reference the consensus definition of contamination, how to clean contaminants from the separator, how to report the findings to appropriate personnel, and whether or not contaminants should be saved for later evaluation. Additionally, they must provide the operator with instructions on how to document the inspection in accordance with the current program.

Operators must be thoroughly trained on the use of the inspection procedure, and the training must be documented. The training record should be documented in a manner that shows a clear traceable record of the qualifications of the individuals conducting inspections on a daily basis.

4.7.4 Cleaning the separator appropriately

With ever-evolving stringent quantitative definitions of what exactly a "clean" surface is, a device that is intended to be beneficial for maintaining food safety may itself be considered detrimental. When a separator is introduced into the process flow, the development of a sanitation standard operating procedure (SSOP) should be considered mandatory. Validation of the efficiency of the SSOP is essential as well. The current state of biological and chemical hazards that are now considered reasonably likely to occur in modern food production necessitates due diligence of equipment sanitation. This is especially true when the separator is in direct contact with the product. Microbiological swabbing, Adenosine Tri-Phosphate (ATP) testing of surface swabs, and residue analysis for specific allergens could provide evidence of the efficacy of the SSOP when such hazards may be present.

4.8 Factors to consider when selecting a magnetic separation method

There are a great number of ways to separate paramagnetic contamination from process streams (as made evident by the previous section of this chapter). Many companies have brochures and catalogs filled with turnkey solutions for the would-be buyer. When deciding on a magnetic separation technique for a specific process, there are several factors that must be considered before engaging in discussions with vendors. In most circumstances reputable vendors will help guide you through this process, but the informed buyer is the easiest to satisfy. Performing upfront homework is wise before calling around for quotes.

Questions that should be answered before selecting a magnetic separator from a vendor for installation within the facility are listed below. By obtaining early answers to these questions, the end result should be a streamlined, successful, and satisfying project implementation.

1. What is/are your contaminants of concern?
    a. What does the hazard analysis display as the contaminants that will be reasonably likely to occur?
    b. What size are they?
    c. What are the requirements of your customer related to these concerns?
    d. Are there any consumer complaint returns that can be analyzed?
2. Location, location, location.
    a. What space is available for the magnetic separation equipment?
    b. Will other process lines, utility drops, or existing brick and mortar need to be relocated to make room for the installation and ongoing service of this device?
    c. Will an employee be able to inspect and clean the magnets housed in this location in a safe and efficient manner? Consider ergonomics as well.
    d. If a suitable location exists, is it in accordance with the needs represented in the hazard analysis?
    e. Could the separator be placed in an area where it would mitigate a hazard and protect expensive equipment downstream?
3. Magnetic separators are not usually very light—this is especially true of plates.
    a. If a magnet is to be suspended over a conveyor, how is it to be anchored in a manner that will not allow it to change position yet still allow cleaning?
    b. Would hangers be used from an overhead, or will brackets be fabricated to secure it? NOTE: C-clamping a magnet to a conveyor or a chute is not an effective or sanitary solution.
    c. Could existing ductwork handle the weight of a plate magnet when under vibration conditions?
    d. Will an actuator be required to swing the magnet on its hinge for inspection and cleaning?

4. How will the necessary exposure of the product to the magnet(s) be maintained?
   a. Are conveyor speeds going to remain constant?
   b. Will the material always be at an appropriate depth, or could there be situations where the material could pile up or avoid the magnet altogether?
   c. How will the control measures be audited?
5. To understand the production rate and material requirements of the process:
   a. Is the amount of product known that will be required by the next process step for an hour of production?
   b. Are there ever surges where demand peaks?
   c. What would be the material throughput requirements if the operation were to increase to maximum efficiency?
   d. What if a 10 percent increase were to be realized beyond what is the current perceived maximum efficiency?
6. Know the characteristics of the material that is to be subjected to the separation technique:
   a. What is the bulk density of the material?
   b. Does the material possess a high viscosity or represent an undue burden for a magnet to draw contamination from?
   c. Will the material flow through the magnetic separator?
   d. What is the particle size of the material, and is it abrasive in nature?
   e. Are the materials corrosive in nature (salometer, pH)?
   f. Does the material readily clump or have random dispersions of lumpiness that could sequester a contaminant?
7. Construction requirements of the separator. Different classes of separation equipment exist: powder grade, utility grade, and food grade. Always specify food grade components when discussing separation options with a vendor.
   a. If gaskets, seals, or lubricants will be needed for the separator, are FDA-approved materials available?
   b. Are they stocked by the vendor?
   c. Is a certificate of conformance issued for each lot?
8. Cleanliness and sanitary design of the separator.
   a. Will the separation equipment be well sealed to prevent leaks or a conduit for environmental contaminants?
   b. Can it be cleaned, rinsed, and sanitized?
   c. Will the separator have a deleterious effect on the flow rate of a clean-in-place (CIP) process?

d. Is there adequate access for dry cleaning if necessary?
e. Will the materials used in the construction of the separator be able to withstand the day-to-day operating environment and provide a long usage life?

9. Formulating a budget—money versus risk? If the contaminants that are to be mitigated via separation are highly paramagnetic and large in size, then the expense of rare-earth materials and multipass sophisticated schemes may not be necessary. Granted, if the product characteristics or contaminant characteristics justify the need for extra expense, then by all means do not skimp. As a point of courtesy be realistic about your budget when speaking with vendors. The vendor knows best as to what level of separator can be obtained in a relative cost range, and can suggest when a stretch upgrade from your initial scope may make the most sense.

One key point to remember is that manufacturing operations are continuously striving to increase their rate of production to better meet demand or to improve their cost structure. Lines that utilize magnetic separators need to be re-examined when production rates increase. The protection that is afforded by the separator will be diminished due to product exposure time to the magnet decreasing, or depth of material increasing. Unfortunately, this undesired effect is nearly in direct proportion to the increase of manufacturing throughput. When installing separation equipment, it is very important that the equipment be sized for the theoretical maximum throughput of the overall operation. If, to save money, a lesser design is implemented, it could result in a costly retrofit later to retain a parity level of FM contaminant separation.

Works Cited and Select Resources

1. Eriez Manufacturing Company. 2000. *Magnetic Drum Separators.* SB–380H Product Brochure. Erie, Pa. 4 p.

2. Eriez Manufacturing Company. 2000. *Magnetic Pulleys.* SB–240N Product Brochure. Erie, Pa. 4 p.

3. Eriez Manufacturing Company. 2001. Ferrous Traps. SB–420T Product Brochure. Erie, Pa. 8 p.

4. Falconer T, Carner M, Norrgran D. *Tips: Understanding How Rare-Earth Magnetic Materials Can Help You Remove Iron Contaminants.* 1977. Powder and Bulk Engineering (5):148–153.

5. Higgins KT. 2002. *Technology Update: Safety in Bulk.* Food Engineering (5):online. Available from: <www.foodengineeringmag.com>. Accessed Feb 03, 2004.

6. Kotz JC, Purcell KF. 1991. *Chemistry and Chemical Reactivity. 2nd ed.* Ft. Worth, Tex.: Saunders College Publishing. 1,066 p.

7. Latimer M. 1989. *Rare-Earth Magnetic Separators: A Cure for Hard-to-Remove Contaminants.* Powder and Bulk Engineering (5):122.

8. Mitra A, Srivastava PK, Bhattacharya DK, Jiles DC. 2004. *Ferromagnetic Properties of Deformation-Induced Martensite Transformation in AISI 304 Stainless Steel.* Metallurgical and Materials Transactions 35A(2):599–607.

9. Miller C. 2000. *Sending Tramp Metal Down the Road.* Powder and Bulk Engineering (12):74–79.

10. *Standard Specifications for Permanent Magnet Materials.* 2000. MMPA Standard No. 0100–00. Chicago, Ill.: Magnetic Materials Producers Association. 28 p.

11. Wilks M. *Six Questions to Help You Select Metal Separators and Detectors.* 2003. Powder and Bulk Engineering (4):34–39.

12. Wilks M, Hueter P. 2001. *How to Select Metal Separators and Detectors.* Powder and Bulk Engineering (12):47–56.

CHAPTER 5

Principles and Strategies for Using Metal Detectors to Isolate Metallic Foreign Materials from Food Products

5.1 Introduction

Use of magnetic separation techniques, as developed in the previous chapter, can greatly reduce the potential for contamination of food products with metals of varying magnetic permeability. The limitations of magnetic separators can effectively be negated when they are used in concert with detection strategies such as x-rays and metal detectors. Metal detectors are ideal for use when the metal contaminant may be imbedded in the product itself or the contaminant is small and not readily reactive in a magnetic field. Essentially, besides x-ray inspection units, metal detectors are the only other commercially available detection method capable of culling nonmagnetic metallic contaminants from a food product. Some examples of the types of metals that are suitable for use of metal detection strategies when intermingled with a food product stream are brass, aluminum, lead, and austenitic stainless steel.

For this reason metal detectors have been used to evaluate food products since the late 1940s, and the proliferation of the use of metal detectors throughout the industry has continued to grow in the last 60 years.

The advent of advanced circuit designs, digital signal processing, and software algorithms now offer advanced modes of operation that were achieved only a decade ago through the use of skilled technicians and attentive operators. The use of metal detection will certainly become more engrained in the food industry as the liberal use of stainless steel equipment expands to address sanitation concerns. An ever-increasing awareness of potential metal physical hazards present in food products has raised the expectation that a better detection strategy be used, an expectation being reinforced in an increasingly demanding nature by the regulatory agencies. Both the USDA and FDA offer documented guidance to their inspection personnel to look for metal detectors and magnetic separators in areas where potential contamination is possible. One point that must be made, however, is that metal detectors are not a true separation strategy. This is because metal detectors rely upon reject mechanisms to be efficient. In essence a metal detection loop consists of three basic components: product transfer, the metal detection system, and a reject system.

5.2 Metal detection—a common CCP within the food industry

When a hazard analysis of a process finds metal contamination reasonably likely to occur despite preventative maintenance and inspection activities, the solution many times is the use of a detection strategy. Usually this will result in the installation of a metal detector or an x-ray inspection system as a catch-all method. The object of installing this equipment is to ensure that the consumer or customer does not receive hard metallic contaminants of sufficient size to be deemed a public health hazard. Most HACCP plans reference metal detectors as the final inspection step of the process to address failure potential at upstream control points. In essence the metal detector is the final step present within the facility to reduce or mitigate the hazard. Figure 5.1 displays a typical convey-through metal detection system broadly used throughout the industry for such a role.

Placing a metal detector on a product line is rarely done to merely address a "quality issue." If microscopic impurities are the processor's main concern, use of various other separation techniques would be recommended in lieu of a metal detector. Metal detection is best saved for identifying contamination that could present a hazard significant to public health. Any third party who walks through a facility will view placement

Figure 5.1. Common convey-through metal detector design used throughout the food industry.

of a metal detector in a process line as a means of protecting public health, not a flag for reacting to process control issues or a method of protecting downstream equipment. Metal detectors can be used for these activities, but realize that the perception of an external auditor will be that the detector is necessary to control a physical hazard.

5.3 General types of metal detectors used in the food industry

Though metal detectors can have vastly different configurations, and are sold by many different corporations, there are essentially only two designs that need to be presented to the reader for identification of potential applications. The two primary types of metal detectors presently suitable for use in the food industry are:

1. Balanced coil designs, also referred to as triple coil type designs
2. Ferrous-in-foil

Single coil designs and eddy current detectors are also available, but their use is quite limited in scope so they will not be discussed further. Each type of detector can deliver good sensitivity and reliability when correctly paired with the appropriate products and environment for operation.

Essentially, metal detectors rely on two characteristics of the contaminants to differentiate them from the product stream:

1. Electrical conductivity
2. Magnetic susceptibility

Product interference factors, to be discussed later, make choosing between the two characteristics for rejection a delicate balancing act. Needless to say, putting a metal detector installation project in the hands of someone who does not understand the drivers and limitations of the technology could be a dangerous move for a processor to undertake. A full understanding of detector types, construction techniques, potential incompatibilities that may be present in the environment, and an intimate knowledge of the characteristic product variation are required to implement a detection program. This chapter is intended as an informative guide to the principles and practices of metal detection applied within the food industry, and it should not be used as a sole guide for purchase, installation, and setup of a detector. Following the guidance of this chapter, coupled with detailed correspondence with potential vendors, will help ensure that the processor chooses the correct detector for the application. When the correct detector is properly installed, the processor should experience a very successful implementation and be assured that there will be a real reduction in risk.

5.3.1 Triple coil designs—typical construction and principle of operation

Bruce Kerr, in 1948, was the first to use a balanced coil design detector for identifying metal contamination in food products, and he has since been awarded multiple patents for the advancement in this technology (1, 2). The formulation of this design was a novel integration of the concepts and technology of induction and detection. Today the balanced coil system is the most commonly used detector in the food industry, and has achieved unprecedented sensitivity and selectivity thanks to advances in manufacturing technologies and interactive digital signal processing. The general principle that governs the function of the balanced coil design has not changed since its inception. It is important, however, that a discussion of the typical components of a balanced coil metal detector be presented first so their function can be better understood at a later point.

There are essentially eight components common to nearly all good balanced coil detector designs used within the food industry:

1. A transmission coil
2. Receiver (search) coils (2)
3. A power supply
4. A basic human machine interface (HMI)
5. A case (rigid)
6. A method of product transfer through the detector
7. A control unit containing the field generator (oscillator), signal metering components, and various output channels
8. A reject mechanism

These components are typically assembled in an arrangement as represented in Figure 5.2. More specific details on the construction of the detector are presented below. This detail may seem mundane, but realize that the sensitivity and specificity capabilities of the detector are determined by the quality of the components used, their geometry upon assembly, and the process controls used in assembling the detector. Adjusting for manufacturing variations in a detector with complex digital signal processing and software compensation, at the point of use, is not a legitimate substitute for a well constructed product.

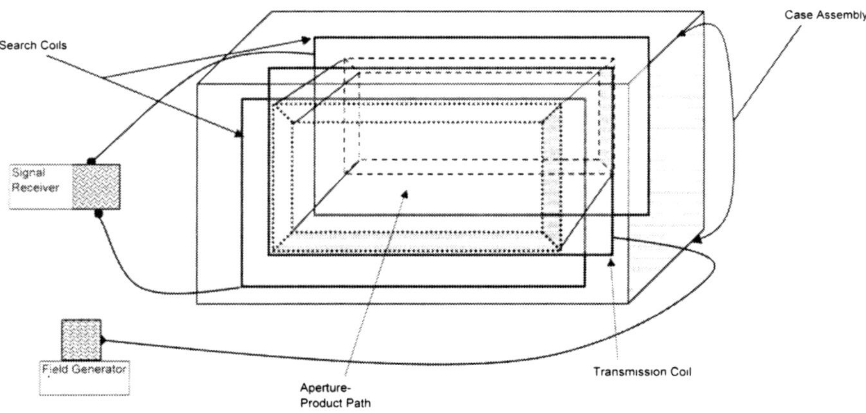

Figure 5.2. Typical components of a balanced coil metal detector.

5.3.1.1 Construction specifics and their effect on the operation of the detector

The materials used in constructing a metal detector combined with the techniques and precision with which they are assembled are the first factors that can affect the performance of the detector. Probably the most critical basic component of any metal detector is the coils that will either transmit or receive the oscillating signal. In order for the detector to work properly, the coils must be constructed within very tight material specifications and maintain tight geometric tolerances compared to one another. This is especially true of the two coils that are going to be used as the search coils within the detector. Advances in digital signal processing have helped to curb the effects of loose tolerances or materials whose properties change over time. To some extent these issues can be dealt with through digital signal processing, but starting with the right materials will go a long way toward ensuring that the detector will function properly as installed.

To ensure that the optimal spatial relationship is maintained between the coils, they will need to be assembled within the case in a very precise manner. To further ensure that the coils do not vibrate or shift, a nonconductive filling material can be introduced into the case as well. The manufacturing process of the detector is such that the two search coils are placed as nearly equidistant to the transmission coil as possible, and the receiving coils are then electronically coupled to the receiving circuit. Keeping the coils within the defined tolerances from one another is absolutely essential for the proper function of the detection unit. A deflection of less than 1 mm from the optimal, equidistant, position could affect the operation of the detector.

The case material used in the construction of the metal detector is also important. This material will need to be rated to handle the adverse environments that the food processing operation will present. Typical adverse environments are wash down, excess dust, and incidental impacts with equipment and personnel. The case must also serve the function of being an electromagnetic shield. It is extremely important to isolate the internal components of the detector from external electromagnetic interference sources—this will ensure the operation is not disrupted by transient interference. Usually aluminum and stainless steel are chosen for the case materials, and these are readily suitable for an operations environment. If the case is to have welded seams, obviously the choice of material used to make the weld needs to be scrutinized, and held consistent throughout the manufacturing process.

The aperture assembly will need to be manufactured from a nonmetal material that is easily cleanable. Usually the apertures of most commercially available metal detectors used in the food industry are manufactured from high-density plastic. Food products may splash or contact the aperture, so it must be readily cleanable and any seams where it interlaces with the case will need to be appropriately sealed to prevent the intrusion of products or cleaning media.

The power supply, control unit, and HMI may be either mounted to the case of the detector (this is known as an integrated design), or linked to the coil circuitry via shielded cable of the appropriate gauge for the installation. As long as care is taken to prevent water intrusion into the shielding cables, or other physical harm (being run over by a fork lift, for example), there should not be a performance difference between the two types. The control panel of the HMI should have a vertical sloping design to facilitate it not being a sink for product or cleaning solutions. The buttons, or keys, on the HMI must be coated or otherwise sealed in a manner that will prevent intrusion of incompatible media as well.

There will have to be a way to convey product through the detector. This can be either in the form of a conveyor belt, ductwork, or piping depending on the material being analyzed. The conveying section that is in the area of the detector must be made of a nonreactive material such as plastic or synthetic fiber. Any imbedded metal pieces in the conveyor mechanism or accumulated static charges can result in false signals or reduced inspection capability due to interference signals being produced. Be sure to check with the manufacturer of the detector to see if there are any specific belting materials, pipe, or chute designs that they would recommend be utilized, or be avoided at all cost.

The reject mechanism that is used in the detector must be one that is of high reliability and robust for the environment. Ineffective reject systems are the weakest link in most detection system designs (5). Several reject mechanisms appropriate for specific utilization in metal detection loops were profiled by Lock (5). Rejection options available for typical metal detection systems include:

1. Air blast
2. Pusher/Punching plungers
3. Sweep arms
4. Diverter arms
5. Conveyor end flaps
6. Retracting belt

7. Reversing belt designs
8. Automatic diversion valves (liquid)
9. Diversion chutes (solids or powders)

It is usually desirable to specify a positive reject tracking system when purchasing a metal detector. This will allow for the detector to track the number of official rejections performed by the rejection device.

5.3.1.2 How balanced coil designs work to detect the presence of metal

The two receiving coils (also called search coils) of a balanced coil detector are balanced in material composition, construction, and spatial relation upon assembly so that they are nearly as exact as possible. When the transmitter coil is energized with an oscillating (AC) current, an electromagnetic field is created that resonates in all directions from the transmission coil. The field can be of varying frequency and its setting will be dependant upon the product, environment, or other interference factors (other detectors in the area), but will generally be between 30 kHz and 700 kHz. Due to a principle known as inductance, the electromagnetic field can cause a current to be produced in a nearby conductor. The con-

Equally spaced on either side of the transmitter coil are two receiver coils.

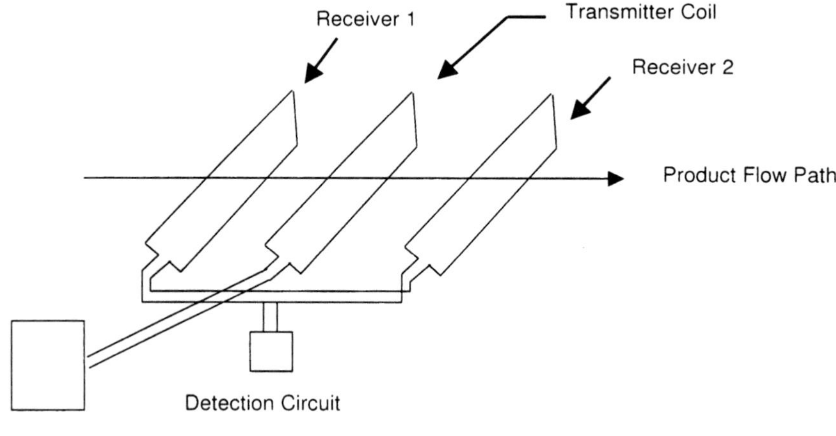

Figure 5.3. Spatial relationship, and interconnectivity of the coil structure in a balanced coil metal detector.

ductors present in this case are the search coils, the product itself, or the potential contaminants. This induction effect is the basis of how generators, radio communications, and broadcast television function.

Because of the spatial relationship imparted during the manufacturing process and diligence in selecting materials utilized for construction of the detector, the receiving coils will be induced by this field in an almost identical manner. The sign of the voltage induced in each coil, however, will be opposite. The receiving coils are connected to one another in a way that nullifies the electrical potential created within the circuit to a zero voltage scenario. Figures 5.3, 5.4, and 5.5 provide graphical illustrations of these concepts. Realize, however, that the signals received will not always be truly negated by one another. This is due to several factors, but the most notable are the variations present within the manufacturing process, the environment that the detector is housed within, and other interference factors that may be transient. Signal conditioning and other digital feedback

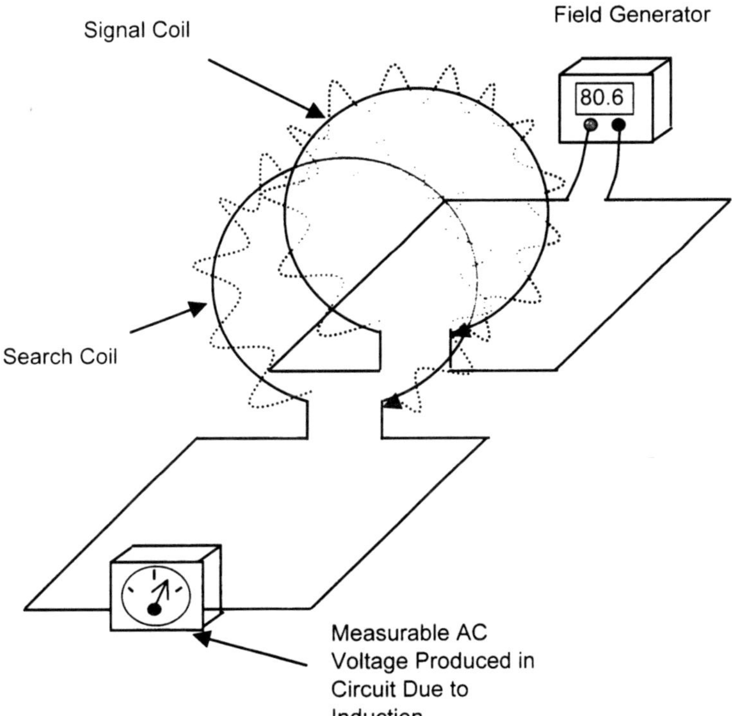

Figure 5.4. Illustration of the concept of mutual induction effect.

Equally spaced on either side of the transmitter coil are two receiver coils.

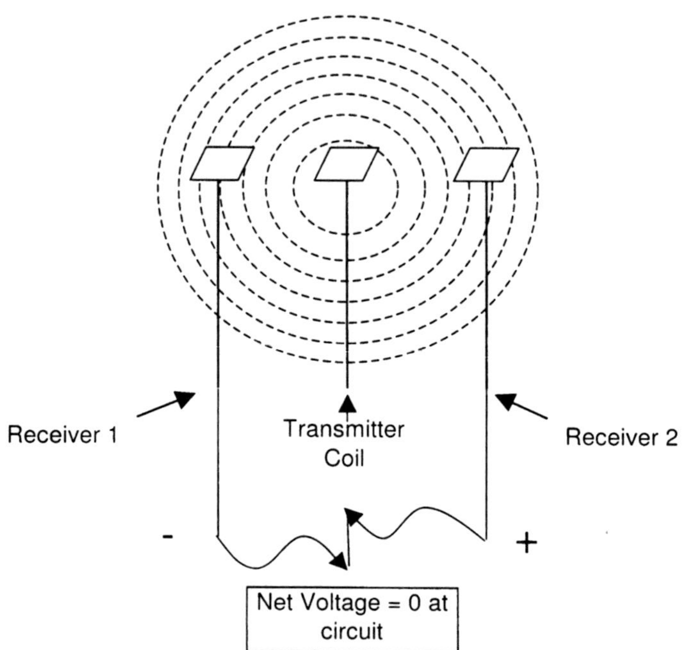

Figure 5.5. Representation of perfectly balanced induction, and the zero voltage situation.

mechanisms that are present in modern detectors can, however, make this issue nearly a moot point.

When metal is introduced to the induction field the balanced coil detector identifies its presence via a loss in the overall balance of the zero voltage state. This is due to the fact that any conductive or magnetic material that enters the field will interact with it. This interaction will produce either an attraction or a repulsion on the field. Both types of effects produced, in the absence of external interference or variable product effects, will be evident at the receiving coils. When this occurs the receiving coils are no longer seeing identical wave characteristics from the received field, and are now unbalanced. This unbalanced state produces a measurable net voltage difference at the monitoring circuit. For many this can be a confusing concept to grasp, but have faith that the unbalanced voltage signal is a repeatable phenomenon. The resultant signal can be used to detect the presence of extremely small pieces of metal within the product. In weakly conductive or nonconductive product types, it can even allow for the characterization of the specific type of metal at the time it is detected.

Amplification of this voltage signal allows for minute changes of smaller than 1/1,000,000 volts to be identified and flagged by the detector. A high-performance radio frequency amplifier can be used for this task, and the output is then modulated down to a lower frequency so the signal can be better converted to a digital format and analyzed. The detector's logic, or circuitry, will key in on the amplitude of the voltage signal, and any shifts in phase, to quickly identify if the unbalanced condition was triggered by metal or if it was caused by other known interference factors.

The signals created by various metals as they pass through the coils of a metal detector can be split into two components: resistive and reactive, according to the conductivity and magnetic permeability of the metal. The signal from ferromagnetic metals, defined in Chapter 4, is primarily reactive to the field; the signal from most stainless steel contaminants found in the food industry is primarily resistive in nature. The signals will vary in amplitude according to the size of the fragment of contaminating metal and in phase (direction) according to the resistive and reactive components that, in turn, depend on the type of metal contaminant (see Figure 5.6).

5.3.1.2.1 Resistive interaction of conductive materials with the detection field

When dealing with induction effects in conductive objects, it is important to understand eddy current effects in conducting materials. Induced eddy currents in these objects flow in what is termed a closed loop manner. Lenz's Law states that, if an induced current flows in a conductor (the metal contaminant in this case), its direction is always such that it will oppose the change of flux that produced it. Stated in a different manner, the polarity of the magnetic field produced in the object that is traveling through the transmitted field of the detector is going to be opposite that of the field from the transmission coil. The eddy current flow, in fact, produces magnetic fields of varying intensity that can interact resistively with the primary magnetic field of the coil. This resistively modified field is

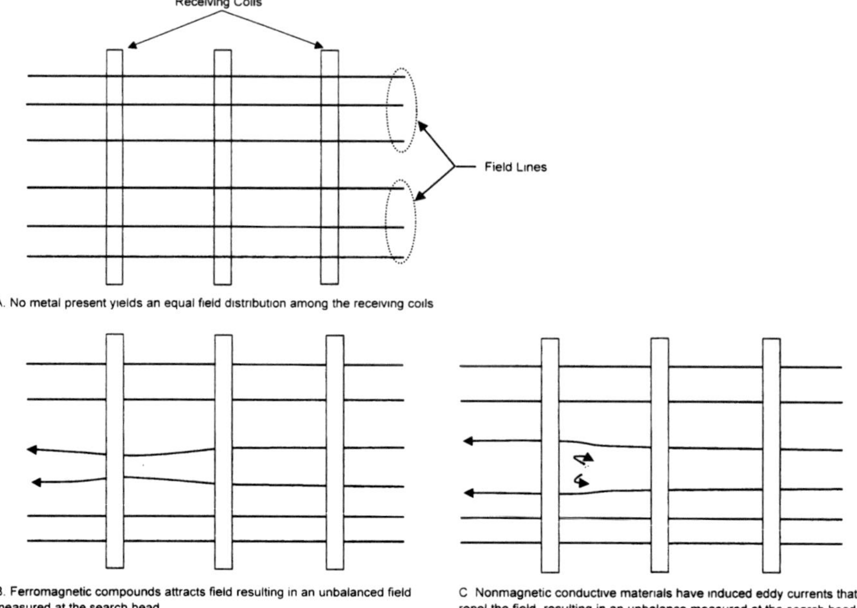

Receiving Coils
Field Lines
A. No metal present yields an equal field distribution among the receiving coils
B. Ferromagnetic compounds attracts field resulting in an unbalanced field measured at the search head
C Nonmagnetic conductive materials have induced eddy currents that repel the field, resulting in an unbalance measured at the search head

Figure 5.6. Field interactions of various metal contaminants.

received at the search head closest to the object, thus causing the induced signal of the search coil receiving the signal to differ from that of its counterpart at the opposite side of the detector. If the resultant difference is determined to be great enough by the detector's signal analysis or threshold value, then a rejection signal would be sent.

5.3.1.2.2 Reactive interactions of ferromagnetic materials with the detection field

Because ferromagnetic materials have free unpaired electrons capable of interaction with a magnetic field, they tend to be well suited for metal detection. The magnetic flux of the induced field is concentrated and strengthened by ferromagnetic materials despite the formation of opposing eddy currents at their surfaces. The increased inductive reactance, due to the inherent magnetic permeability of these materials, makes it easy to distinguish them from food products that are passing through the detector. The signals presented by ferrous compounds are so strong that extremely small contaminants (< 0.8 mm) can be reliably detected in many applications.

5.3.1.3 Factors that can affect the sensitivity of the detector and suitability for detection

The sensitivity of a given detector will vary based on its installation, use, and age. There are also several basic drivers that can make the sensitivity of a detector vary with respect to specific contaminants. These must be fully understood by the processor before attempting to address a metal contamination issue. As will be discussed below, some product and packaging materials may be totally incompatible with a balanced coil metal detector. For this reason it is best that the processor have present a thorough description of the product to be inspected, as well as various product samples, when discussing metal detection options with a vendor.

5.3.1.3.1 Type of metal contaminant to be detected

Obviously the type of material that the contaminant of interest is composed of has a big effect on the sensitivity achievable. Detection of ferromagnetic materials is performed with ease when a properly functioning detector is placed in a suitable environment. The detection of nonmagnetic conductive metals like lead and copper is usually also not an issue as long as the conductivity of the product does not mask their presence from the detector.

The detection of stainless steels, however, that are neither good conductors nor paramagnetic in nature (304 and 316 with a highly austenitic nature) can be much more difficult even with digital signal processing algorithms. It is widely understood throughout the industry that the sensitivity of a detector to stainless steels when compared to ferrous materials is at best approximately a 1.5:1 relationship. This means that the measurable signal presented to the detector from a 1.5 mm stainless steel test sphere is equivalent to that of a 1 mm ferrous test sphere. In reality, when wet conductive products are involved, the sensitivity can decrease to > a 2:1 relationship.

The frequency of the transmission signal can help augment signals from different types of contaminants. Increasing the frequency of the transmission signal will increase the signal received for all types of metals in a dry nonconductive product environment. High frequencies help to induce stronger magnetic moments in ferrous materials and increase the eddy current formation and signal resistance in other materials. It is advantageous to run a detector at the highest frequency that affords differentiation of the potential contaminant from the product without inducing false rejection signals.

5.3.1.3.2 Detector design and aperture size

The physical dimensions of the coils are one of the key factors to determining sensitivity to conductive and ferromagnetic materials. A large internal coil area will result in decreased sensing ability across the aperture. Regardless of size, the least sensitive point of any balanced coil metal detector is at its geometric center. Whenever a detector is purchased it is extremely important that its sensitivity be specified at the center of the detector. Beware of misleading advertising in this regard. Figure 5.7 is a representation displaying the size of a sphere necessary to generate the same contamination reject signal at various conveying points through the detector. Note the size of sphere necessary to set this specific detector into a reject mode at its center, then compare the object in the center with those at the corner and sides, respectively. Note that the advertised sensitivity of this detector is really based on an average of these, and does not represent the sensitivity of the center area. This factor is easy to understand when one thinks about the fact that the generated electromagnetic fields, from both the transmitting coil and subsequently the product itself, will follow the inverse square rule in regard to loss of field strength. Induced magnetic fields and eddy currents within small particles would require a sensing

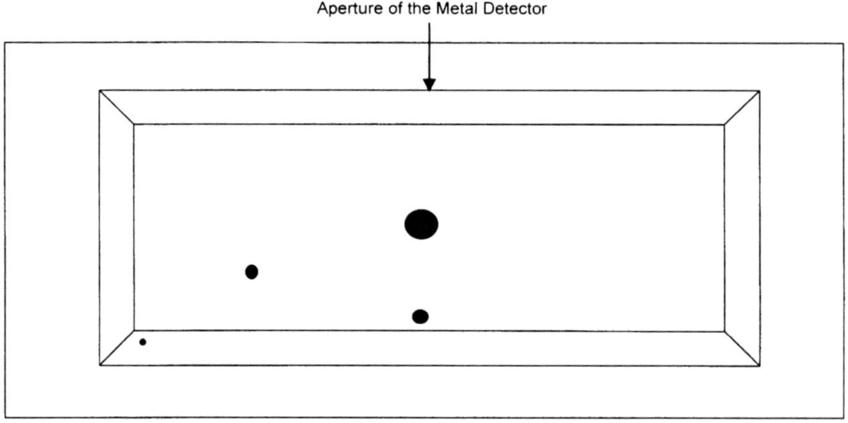

● = "Advertised" sensitivity of the metal detector (not specified at the center of the aperature)

Figure 5.7. Representation of varying sensitivities present as products are conveyed through a metal detector at differing locations. Note that the advertised sensitivity of the detector is not a true representation of the detector's weakest sensitivity areas.

coil to be in a very proximate relationship to it for detection via signal interference. An often quoted rule of thumb is that normal sensitivity of a detector can be anticipated to be about one percent of the coil diameter. Realize, however, that this is a statement for equidistant coils being square or circular in shape. For rectangular applications, if the width of the coils is greater than the height of the coils, a further reduction in sensitivity can be experienced.

5.3.1.3.3 Product effects

The product itself can have a deleterious effect on the sensitivity of the detector. Other than dry ingredients and certain oil blends, most food products are conductive in nature. Granted that you would not try to use them as a substitute for a broken cable, wire, or antenna, their inherent conductivity can still affect the operation of a metal detector. Products containing a water-based component and high quantities of ionic salts (pickles and condiments, for example) present the worst possible interference for the function of a metal detector. Products with water on their surfaces can also affect the received signal of the metal detector. The reason for this effect is because water, in the presence of minerals, is an excellent conductor of electricity. The transmitted signal from the detector produces eddy currents at the surface of the products, just like it would with any other moderately conductive metal. The resultant signal interference can be identified at the receiving coils, and makes it difficult to detect certain conductive metal contaminants.

To deal with such an issue, many manufacturers have incorporated digital signal processing and feedback control mechanisms. Essentially, the detector learns the variation presented by the product in digital form, then sends an analog signal back to the receiving coils to balance out the product effect. The algorithm used in the logic of the detector may even automatically adjust the detector's frequency to optimize the received signal for detection. Most of the modern detector styles compensate for and learn the signal slowly, and have become adaptive to certain amounts of variation in the product (1, 2, and 3).

5.3.1.3.4 Product effects 2—packaging materials

Product effects are not limited to the consumable components of a food product. Variation within packaging materials and types of materials used in their structures can present issues for detection as well. When inspecting a product that is packed within a heavy corrugate case, expect a number of false rejects. These rejects truly are not false, however, because the detector will be keying in on metal contaminants commonly found in the die cut and scored fiberboards. Finished-case inspection appears very attractive for providing assurance that the products are free of defects as they leave the facility. This method can also be attractive because it can potentially limit the number of detectors required, but due to the packaging issues above, its use should be carefully considered.

Obviously, using metal detectors to inspect cans of soup or products packaged in glass jars with metal lids is not a suitable or reliable application. Certainly the edible portion of these products can be conveyed through a metal detector prior to filling, but using a metal detector on the finished package itself is not feasible due to interference. Products encased in metal or foil film packaging materials do represent a challenge for detecting contaminants, but they too can now be inspected.

5.3.1.3.5 Orientation effects

Nonspherical contaminants and products will deliver a signal that varies according to their specific orientation (5). Metal wire, swarf, and conductive products like sausages can all present an orientation effect to the strength of signal received by the metal detector. The orientation effect for a contaminant will only be a concern if the contaminant is smaller than the minimum spherical particle detectable of the same material. Stated simply, if the metal detector can reliably detect a 1.5 mm spherical particle of stainless steel at the center of its aperture, then a 1.5 mm diameter stainless steel wire will not present an issue for detection if its aspect ratio is >1:1.

Where wire contaminants can become problematic and potentially the source of a market action is when they are smaller in diameter than the minimum particle size detectable by the detector. Each wire contaminant has an optimal and a worse-case orientation that it can assume as it passes through the detector. The orientation for optimal or worst case detection is based on the composition of the wire. The orientation effects for each type of wire are depicted in Figure 5.8. Realize, however, that rigid wire is considered a physical hazard by FDA and FSIS, regardless of its diameter, if it is greater than 7 mm in length.

5.3.1.3.6 Other interference factors

The environment that the detector is placed within can be a reason for poor operation and reduced sensitivity. Ideally the detector will be placed in a manner where the operators can have free access to it to ensure that it

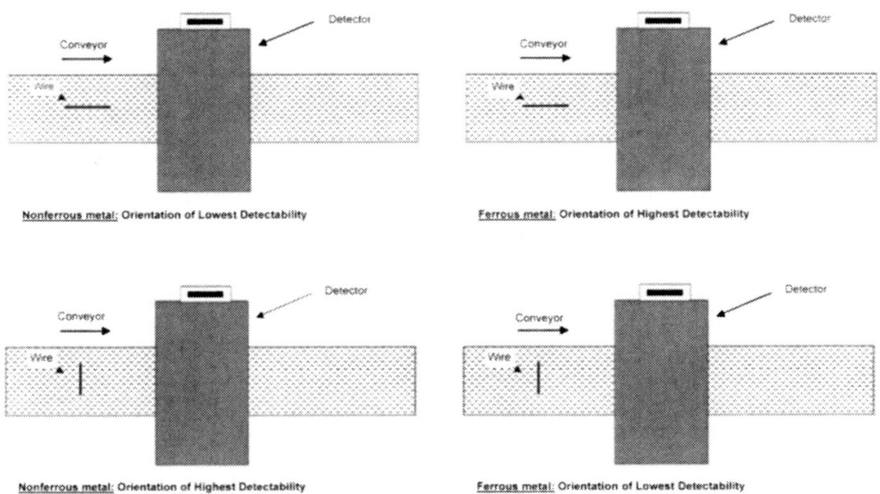

Figure 5.8. Orientation effects of wire contaminants based upon their composition.

is functioning. This area must be designated as a metal-free zone. The sensitivity area of a detector goes beyond just the area inside of the aperture; as a rule, the space on each side of the aperture should be free from metal parts and equipment with moving metal-to-metal contact surfaces (rollers, loosely bolted supports, etc.) or conductive lubricants (5). Painting a visible marker on the floor around the detector and attaching plastic rods of the same color on top of the detector are good practices to remind folks in the area of the spatial requirements of the metal-free zone. This area should be twice the height of the aperture to ensure the detection field is not corrupted (5). By placing visible cues in the area of the detector, it becomes very easy to audit the area for compliance. Objects commonly found to violate the metal-free zone are clipboards, various metallic personal effects of employees, and hand tools.

If multiple detectors are to be used within the same building, they can be a source of interference for one another. Usually the detector manufacturer can provide information about the potential effect that may be presented by this situation, and even remedy the issue in the design phase before it becomes a problem. Vibration of equipment and the quality of the electricity fed to the detector can affect the performance of the detector as well. Modern digital detectors have come a long way in ensuring that they are more robust in such climates, but due care should be taken upon installation and setup to ensure that these factors do not become a problem

Figure 5.9. Illustration of a possible source of environmental interference in a metal detection scenario.

later. Also realize that placement of the reject device and the path of any electrical conduit can have an undesirable effect on a metal detector. The unshielded drop cord presented in Figure 5.9 would be a likely source of interference for the detector if energized, due to its proximity to the aperture.

5.3.2 Ferrous-in-foil metal detectors

The proliferation of laminated packaging materials incorporating metal-impregnated films that serve as barrier layers has presented a significant challenge for the balanced coil detector. If the film is conductive in nature, it is possible to tune out the effects of the metal-impregnated layer while still using a balanced coil design. The sensitivity in this case will be dependent upon the quantity and size of the contaminant present.

To achieve sensitivity for contaminants in metal-impregnated film packaging, a very low frequency must be sent through the transmission coil, and the detector's sensitivity to small contaminants of all types will be markedly reduced (usually by twofold or more).

For this reason, balanced coil design application to operations handling metal-impregnated films is a risky option, especially when the processor is already trying to deal with small contaminants or wires. X-ray systems may present a more appropriate option for this situation. If ferrous contaminants are the major concern, however, and the metal in the film is nonreactive to a magnetic field (aluminum, for example), a ferrous-in-foil-detector design would be an appropriate choice. The limitation of this detector type is that it can only detect ferrous metals and ferrous alloys that are highly reactive within a magnetic field.

The ferrous-in-foil detector is akin to the magnetic separator, except that it does not rely upon magnets to pull the metal contaminant from the product burden. Instead it senses the presence of the metal through magnetic induction, and flags a reject mechanism to dispose of the product which contains the contaminant. Ferromagnetic contaminants and contaminants with increased magnetic permeabilities (including martenistic stainless steels) can be reliably detected by this type of detector. The strength of the magnets utilized in the construction of the detector and their orientation are key for the operation of the detector. The distance that the search coil arrangement is from the contaminant will also represent a critical factor that affects the sensitivity of the unit.

For this reason the aperture heights for these detectors is quite limited. Typically the only adjustment that can be made on these types of detectors is the sensitivity adjustment, but manufacturers are starting to use digital signal processing techniques to make the units more robust. The construction of a typical ferrous-in-foil detector is presented in Figure 5.10. Use of stronger rare-earth magnets or electromagnets in the construction of these units would surely help increase the sensitivity and size constraints; however, such designs are not practical because of their cost.

5.4 Handling detector rejects

Under no circumstances can it be assumed that a product rejected from a metal detector is contaminant-free and can be returned to the line without further examination. Policies and procedures dealing with the handling of, and reinspection of, rejected product must be clearly defined by

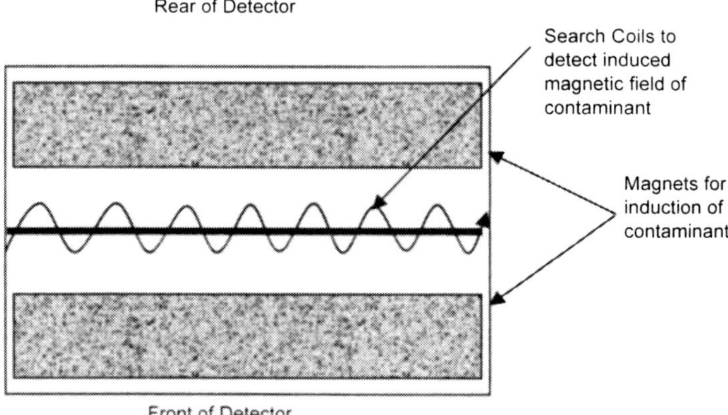

Typical construction of a ferrous-in-foil metal detector- top of case omitted

Figure 5.10. Typical construction of a ferrous-in-foil metal detector relying on magnetic induction versus eddy current sensing.

the processor. Personnel who work around the detector, and those who maintain it, must possess explicit knowledge of these documents and demonstrate that they understand the intent of the policy. In loosely managed environments the reintroduction of contaminated product to a process line, at a point past the detector, is a real possibility.

For this reason rejects should be reintroduced to the detection process only during periods of down time, and the real false rejects must be accounted for. Due to position effects of contaminants it could be possible to have a contaminant initially be rejected by a detector, only to pass a second detection. All rejected items must be passed through the detector multiple times. A minimum of three nonrejecting passes conducted at differing orientations to the detector is minimum due diligence. Any sample that is rejected in this exercise will need to be mined for metal to better understand the potential severity and scope of the contamination.

Inspection of the true rejects should be performed using the same detector, if possible. The detector should be utilized as the main tool for the investigation because it is far more sensitive to metal, potentially occluded in product, than the human senses of sight or touch. The object should be presented to the detector in subcomponent form to offer the best chance of finding any small pieces of metal. Obviously opening the product container and finding large metal objects present, like a sample scoop or other utensil, would probably negate the necessity of the following exercise, but otherwise continue as presented below:

1. Remove the item from its packaging (if applicable), and send the packaging and product separately through the detector. The object should be placed on the detector so that it will pass toward the sides of the aperture to optimize sensitivity.
2. If only the packaging material sets off the detector, then examine it to determine the type of contamination and its potential source. Otherwise, begin to subdivide the product into smaller fractions.
3. Run the product fractions through the detector until the metal object is isolated.
4. Examine the metal object and note its visual characteristics (flake, wire, swarf, etc.). Use an Alnico bar magnet for a rough determination of its composition.
5. If the object is readily identifiable and its source is known, proceed to root-cause identification while dealing with the suspect product (issue a hold, cease production, clean equipment, reinspect product, etc.).
6. If the object is not identifiable and no source is known, prepare for a lengthy investigation and begin the process of dealing with any suspect product.

5.5 Recordkeeping

It is imperative to document the function of the detector and any products that may have been rejected by the detector. Possessing proper documentation of detector functionality can aid the processor when processes are reviewed by regulatory agencies, lawyers, and/or customers. The records kept should detail what test spheres were utilized to challenge the detector and the result of each test performed. Obviously other factors like the time and date of the test, the product being run, and the initials of the operator who performed the test are also advantageous to have on the record. When documenting the function of the detector, use the affirmative statements "YES" and "NO" instead of the sometimes legally ambiguous "OK." Any corrective actions taken for a failed test must be recorded on the record as well.

A second record should be kept that deals with the number of rejects or false rejects for the specific detector. This information allows the processor to possess a snapshot of the day-to-day operation of the detector. The data can also be potentially used during record review to withhold production runs from distribution for being over a certain reject threshold (potentially indicating that there could be a larger problem). Over time this data can become quite useful for many other purposes. Chapter 9 describes techniques on how to trend this data to better understand any undue variation that may be present.

Works Cited and Select Resources

1. Kerr B, inventor; Goring Kerr Limited, assignee. 1986 Jan 7. Inductively balanced oscillatory coil current for metal detection. Submission for Patent No. 4563645, United States Patent and Trademark Office.

2. Kerr B, inventor; PLC, assignee. 1988 Jan 12. Metal detector for detecting product impurities. Submission for Patent No. 4719421, United States Patent and Trademark Office.

3. King KJ, inventor; Ramsey Technology, Inc., assignee. 1997 Nov 25. Forced balance metal detector. Submission for Patent No. 5691640, United States Patent and Trademark Office.

4. Lock AP. 1996. *Tips: Evaluating Product Rejected by Your Metal Detector.* Powder and Bulk Engineering (6):72

5. Lock AP. 2002. *Metal Detection.* In: Food Plant Sanitation. Pederson, N.Y.: Marcel Dekker, Inc. 752 p.

6. Wilks M. 2003. *Six Questions to Help You Select Metal Separators and Detectors.* Powder and Bulk Engineering (4):34–39.

7. Wilks M, Hueter P. 2001. *How to Select Metal Separators and Detectors.* Powder and Bulk Engineering (12):47–56.

CHAPTER 6

Machine Vision and Its Application to Prevent FM Contamination of Foods

6.1 Introduction to machine vision

Many processors utilize machine vision to automate routine product inspection activities. Machine vision defined by the Automated Imaging Association (AIA) is "The automatic acquisition and analysis of images to obtain desired data for controlling a specific activity." The beauty of machine vision is that it is an ever-developing technology that offers more functionality as time passes. Essentially, if the processor has enough resources, a machine vision strategy can be employed at practically any point in the process that a human being is currently stationed to examine objects by their color, shape, or other visual cues. Current applications of machine vision systems in the food industry range from checking the distribution of icing on a toaster pastry to ensuring cranberries are not intermingled with deer feces. There are many advantages to utilizing a machine vision strategy instead of manual inspections for such activities, the most notable of which are:

1. Machine vision systems can accommodate extremely high throughputs due to the efficiency of their microprocessor components.
2. Machine vision systems do not suffer from boredom, fatigue, employee relations issues, or require breaks for lunch.

3. Machine vision systems, once "trained," offer high degrees of sensitivity and can be used to see spectra that humans cannot (InfraRed, UltraViolet, X-ray, Ultrasound, etc.).
4. Machine vision systems represent a high initial investment, but usually do not require substantial maintenance expenditures to keep them reliable.
5. Machine vision systems can be used to provide the feedback component within a control loop for an automated process such as band ovens or peelers.

Obviously the disadvantages of using a machine vision system are rooted in the fact that they are truly not intelligent, as are humans. Some of the more complicated neural networks and fuzzy logic schemes that have been developed for these systems allow for minor cognition through multiple mathematical operations. This cognition, however, cannot begin to approach the human intellect. A machine vision system, for instance, would see speckles of food-grade grease on the surface of a product as a "defect" and promptly reject it if the right threshold and morphology recognition software algorithms were used. A human, on the other hand, may spot the grease and make the instant connection of, "Wow, that's grease! We'd better shut down the machine and figure out where it's coming from."

6.2 How does a machine vision system work?

In order to fully understand how to apply a machine vision strategy to prevent foreign material contamination of foods, one must first understand the basic principles that govern their function. From this point it will be much easier for the reader to understand how the components of a typical machine vision system interact with one another to identify foreign materials and subsequently reject them from the process. The typical components of a machine vision system are:

1. A lighting source to illuminate the inspection area
2. A photo-sensing device
3. A product presentation scheme
4. A frame grabber that serves as an analog-to-digital converter

5. A computer programmed with software and any algorithms necessary, as well as input/output (IO), for a reject signal
6. A reject mechanism

6.2.1 The electromagnetic spectrum and visible light

All foreign material detection strategies used in the food industry utilize regions of the electromagnetic spectrum (see Figure 6.1) to differentiate contaminants from product. Each type of electromagnetic radiation possesses characteristically different energy levels. This energy can be imparted on any object that is in the path of the source of the electromagnetic radiation. This fact is a true statement whether the electromagnetic radiation present is of low frequency (like a metal detector) or extremely high frequency (like an x-ray source). The attenuation of the spectrum is

Visible Spectrum

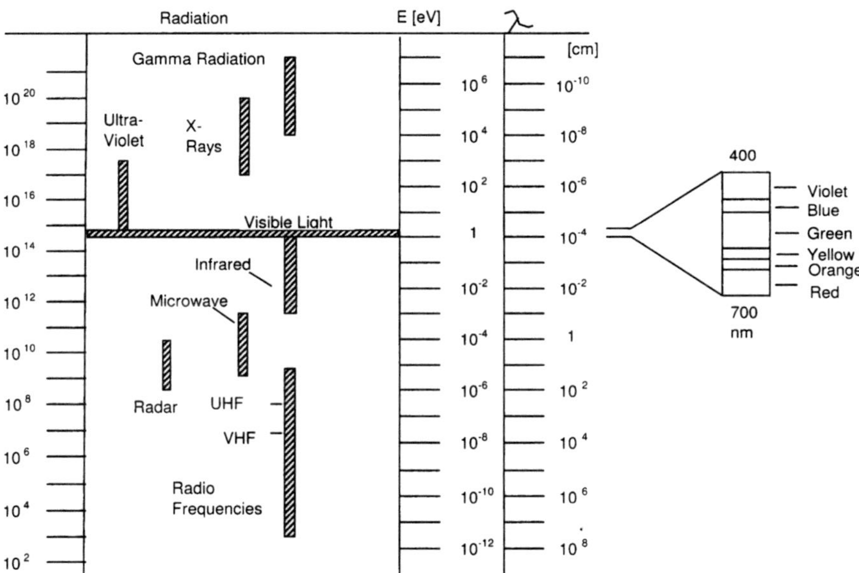

Figure 6.1. Electromagnetic spectrum. Note that visible light is between 400 to 700 nanometers.

basically isolated and quantified by a detection system to reveal the contaminants. Machine vision systems that use visible and near-visible light will be the primary focus of this chapter. Modern automated x-ray inspection is also a form of machine vision, but due to its widespread use and several other nuances, it will be covered in detail within the next chapter of the text.

The important concept that must be grasped is that visible light in itself is a form of electromagnetic radiation. All electromagnetic radiation can exist simultaneously in two forms: a wave and a particle. Thanks to the work performed by physicists in the late 19th century (namely Maxwell and Planck), it was discovered that electromagnetic radiation was present in a wave form possessing a wavelength, frequency, and amplitude. The quantum theory was also born during this period when Planck developed an equation to relate the minimum emitted energy (Planck called this a quanta) of electromagnetic radiation proportional to its frequency.

It wasn't until 1905 that Albert Einstein was able to refine the existing theories to demystify a then poorly understood phenomenon known as the photoelectric effect. The photoelectric effect baffled physicists because experiments had shown that, when certain metals were exposed to light, possessing a minimum frequency, the metals would eject electrons. This electron flow was captured by placing a positive electrode in a vacuum tube to attract the negatively charged electrons that were being ejected by the metal. Attaching a crude analog detection meter in the circuit made between the metal and the electrode outside of the tube allowed for quantification of the electrons and their respective energy (3). The physicists noted that, when the intensity of the light was made brighter (increased), the number of electrons ejected from the metal increased, but the energy measured in the circuit remained the same. The interesting part was that when the scientists decreased the frequency of the light source below a minimum threshold, the entire electron flow abruptly ceased, but increasing the frequency of the light source caused the electrons to flow again. Further increasing the frequency caused the flowing electrons to posses more energy as they were ejected from the source.

This phenomenon could not be explained by the current paradigm that light was a wave. Einstein came to the unprecedented conclusion that a beam of light has wavelike properties, yet it is also a stream of particles which he termed photons. Einstein deduced that the photons of the light that were hitting the metal possessed enough energy to cause the electrons to be ejected from it. He also theorized that the energy of the photons increased in proportion to the frequency of the light source, and this fact explained why the electrons that were ejected from the metal object possessed higher energies. Simply, the photons would impart their excess energy into the electrons causing them to catapult away from the metal with increased energy. This can be likened to a frame of bowling, where a weak person throws a ball down the lane with only enough energy to cause one pin to wobble and fall over while not disturbing the other pins. The energy imparted on the pin via the ball is not enough for it to perform work on any of the other pins—conversely, when the same ball is thrown by a different person with a much higher force and hits the same pin, it will be sent flying with excess energy. In fact, there may be enough energy imparted in that pin for it to knock over other pins in its path.

The equation that Einstein developed to describe this phenomenon was $E = h*\gamma$, where E is equal to the energy of the produced electron, h is Planck's constant ($6.63*10^\wedge - 34$ J*s), and γ is the frequency of the light source. This equation mathematically demonstrates that the energy of a photon is not dependent upon the intensity of the light, but rather it is directly proportional to the frequency of the light source. Einstein also stated that the intensity of the light (brightness) was also related to photons. In this case, however, higher-intensity light meant more photons being generated by the source. These excess photons created the brighter light observed by the physicists. This is why increasing the intensity of the light source would increase the number of electrons ejected from the metal, but would not affect the measurable energies they possessed. Going back to the bowling example discussed earlier, an increase in light intensity would be like throwing multiple balls down the lane with the same force. If the force applied to the balls was sufficient, and they were aimed correctly, they could each knock over a pin.

The wavelengths of visible light apparent to the human eye begin with the perception of the color red in a range of 622 to 780 nanometers. The colors that follow in order of decreasing wavelength (meaning a subsequent increase in frequency) are: orange (597 to 622 nanometers), yellow (577 to 597 nanometers), green (492 to 577 nanometers), blue (455 to 492 nanometers), and finally shortwave-length violet (455 nanometers and below). The infrared (IR) region lies at a longer wavelength (shorter frequency) than the visible red region, therefore infrared sources produce photons of lower energy. Conversely ultraviolet (UV) light exists at shorter wavelengths (higher frequency) than the visible spectrum, and produce photons possessing higher energy. The increased frequency (or energy) of UV light sources results in photon energies that can damage skin cells and can be used to destroy microorganisms present on nonporous surfaces and in clear liquids.

6.2.2 What is sight?

Before illustrating how a machine can see, it is probably best that human sight be introduced first. For a human to see an object, the first prerequisite is that their eyes have to be open. After accomplishing that milestone there must be adequate lighting for the eye to function, and the object of interest must be presented in a manner that lends itself to being visible (i.e., not an extreme distance away, not extremely close, and not shadowed). When these conditions are met the light that originated from the lighting source (sun, light bulb, candle, light-emitting diode [LED], etc.) is reflected by the object in a divergent manner. These divergent waves of light of varying frequencies will pass through the eye's optical system (the lens and the cornea), which focuses the image upon the retina.

Due to the curvature of the cornea the image projected and focused on the retina is now upside down and reversed right to left from the original image (see Figure 6.2). The light that is focused on the retina must pass through a thin layer of tissue consisting of cells and nerves before it reaches a series of photoreceptors. The photoreceptor cells of the retina, known as cones and rods, react to light through a series of electrochemical reactions caused via absorption of light photons (this works in a similar manner to the photoelectric effect). There are three types of cone cells present in the human eye, and each type contains a unique photopigment that is stimulated differently by photons of reddish-yellow wavelengths, green wavelengths, and blue wavelengths, respectively. These cone cells afford the differentiation of color in the visible spectrum by human beings. Electrical impulses created from photon absorption at the cones and rods

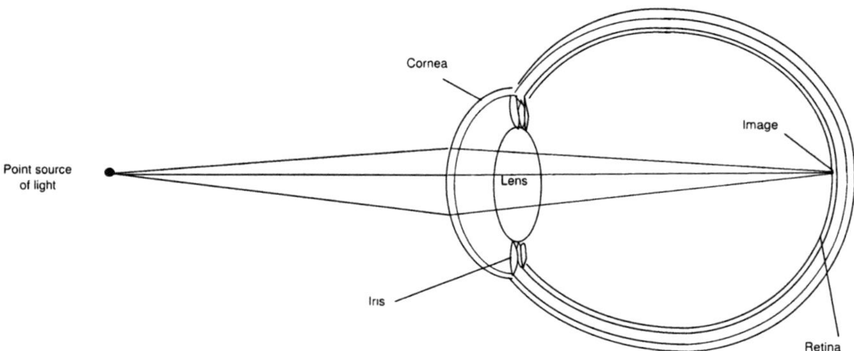

Figure 6.2. Functional diagram of the human eye.

will subsequently travel through the optic nerve to the brain, where they will be processed and constructed into an image. As a side note, individuals who are color-blind usually lack certain photopigments; therefore, their images will be lacking in certain colors when processed by the brain. All of the colors of visible light can be represented with the three colors red, green, and blue (RGB). Figure 6.3 displays the colors of the visible spectrum that are perceived by the human eye. Note that the colors cyan, yellow, and magenta (CYM) are obtained through subtractive interactions of RGB.

Comprehending how humans function to see FM components in food products or subcomponents makes it very easy to understand the basic requirements that need to be met for a machine to "see" the FM as well. Obviously one would not expect a human inspector to identify chunks of wood or plastic present in a product stream when the lights are turned off. Likewise, posting an inspector at a workstation in a manner where they are able to observe only a small portion of the process stream would not be effective. Another inadequate situation can be found when a human being is

RGB Color Space

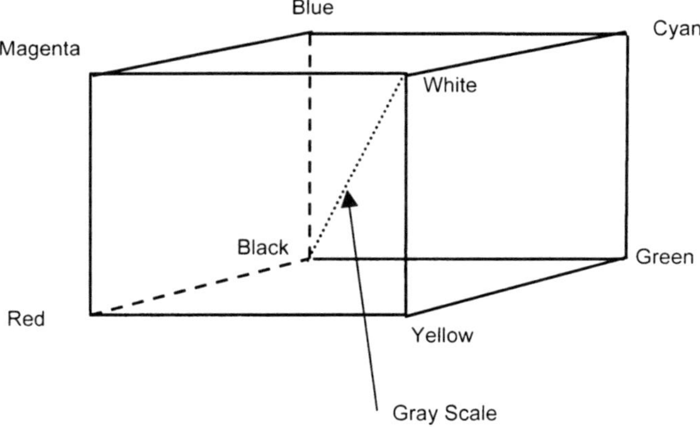

Figure 6.3. Drawing of the color cube. Note where the gray scale is represented.

asked to sort out defects from a material that has a heavy burden or depth. This situation doesn't facilitate the inspector being able to identify contaminants below the immediate surface presented. Some may laugh at such common sense examples to illustrate proper application of machine vision technology, but many companies spend exorbitant amounts of money trying to remedy errors that originated from fundamental design assumptions such as these. Many times the root cause of such errors is failing to expend effort and resources on upfront planning for the requirements of the system.

6.2.3 Lighting

Just like the human eye, machine vision systems need adequate lighting to function and the choice of the light source used will be based on the features and/or wavelengths that are of interest. One must realize that not all light sources are created equal. A standard filament bulb, for example, does not emit every wavelength of the visible electromagnetic spectrum in a uniform manner. This varying distribution of wavelengths can affect the light reflection from the object being imaged. For this reason light sources must be chosen carefully.

Typically the sources available for machine vision systems are incandescent, halogen, fluorescent, or they may originate from arrays of light-emitting diodes (LEDs). Each type has advantages and disadvantages associated with their use, so it is usually best to work with the vision system manufacturer or retained integrator to choose a light source that is correct for the application. It is also important to realize that any unplanned fluctuations in external lighting may "pollute" the desired lighting scheme and result in the vision system not functioning properly. Any light source chosen, however, will need to be adequately shielded in a manner that prevents it from becoming FM itself. Placement of the light source is also important because, if it is in an area where splatter or soiling may occur, it can possibly fail to provide adequate lighting levels for product inspection, resulting in dead spots or contrast issues. All light sources must also be located in a manner that offers easy access by maintenance personnel to check and change the source if necessary.

Once a source of the desired characteristics has been selected, it is time to decide how to best illuminate the product to be inspected. Machine vision systems, like human vision systems, are affected by shadows and bright spots. The approaches to illuminating the inspection objects in machine vision scenarios vary via the specific objective, but can be classified into two general approaches: direct lighting of the object or diffuse lighting of the object. Direct lighting of the object is simply that the light source is shined directly at the inspection area. Diffuse lighting is when the light source is directed toward a diffuser material (typically a plastic screen) that emits photons of light in multiple diffuse paths. Typically, direct lighting tactics will yield more shadowing than diffused light. The light type is important, but equally important is the fact that it can be introduced in several different geometric relationships relative to the sample and the "eye" of the vision system, depending on the result desired.

Backlighting, where the path of the product lies between the light source and the eye, is commonly used to determine the shapes and overall sizes of objects present. Essentially, the vision system would recognize the object in this scenario via a noticeable blockage of the light source. Obviously this type of application offers little opportunity for the vision system to discriminate defects or variations that may be present at the surface of the object. For situations where this is important, directional lighting using either a direct or diffused light source would be the better choice. This scenario would enable the eye to receive transmitted light from the surface of the object as well as any adjacent unassociated object forms or background materials. Because the light source is directional, reflections of transmitted light from an irregular surface will be made more apparent to the vision system due to scattering. With proper use of multiple directional lighting setups, multiple vision systems can be arranged in such a manner that each is analyzing the same field of view from different angles. Figure 6.4 illustrates such a system. This information can be combined to generate nearly 360° views of products. Certain other lighting schemes can also provide depth of field to achieve information in three dimensions.

When a technique known as structured lighting is used, depth features of the object can be seen as well as the two-dimensional factors already illustrated. A typical structured lighting application would be to project a light grid onto the inspection area. Objects passing through the grid will cause a visible distortion of the light that is a variant from the established empty cell grid previously presented to the vision system. By measuring these discrepancies, additional shape features may be extracted by the vision system to separate good products from FM.

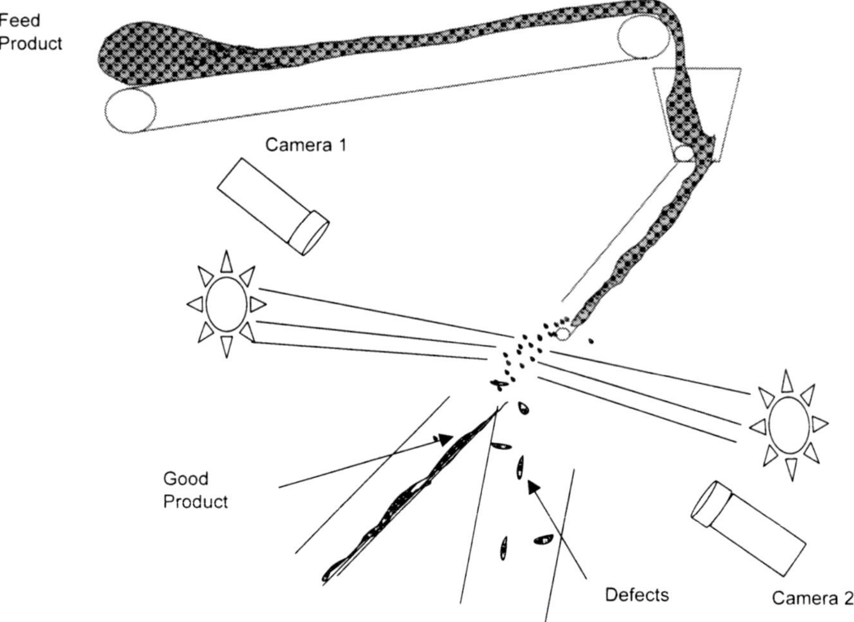

Figure 6.4. Lighting and cameras necessary for a near 360° field of view with a machine vision system.

6.2.4 Photosensors for machine vision systems

Choosing a lighting strategy is only one part of the equation necessary for constructing a functioning vision system. The system will need a set of "eyes" to be able to sense the wavelengths of light that are characteristic of the objects of interest. There are several photosensing components that can be used to process electromagnetic radiation into electrical inputs for processing via machine vision systems. The most commonly used sensors for contemporary machine vision systems are:

1. Linear arrays of photodiodes
2. Change coupled devices (CCD) line scan cameras
3. CCD area detectors
4. Complementary metal oxide semiconductor (CMOS) area detectors

Each of these will be explained below, but it is important to realize that all of these sensors rely on photons impacting a unit area to produce a measurable flow of electrons. With the ever-decreasing cost and increasing reliability of manufactured semiconductor components, these key components of vision systems have reached unprecedented levels of high performance, accompanied by lower pricing.

6.2.4.1 Semiconductor principles introduced via the photodiode

The photodiode currently represents the base level technology for image acquisition in machine vision systems. A photodiode is essentially an assembly of two electrodes formed in a single crystal silicon wafer. This wafer assembly is similar to those used in the manufacture of integrated circuits, and photodiodes are commercially available from numerous manufacturers. Figure 6.5 displays a photo of a common 0.8 mm diameter photodiode. Figure 6.6 is a cross-sectional representation of the elements that the photodiode consists of, and also displays the photon interactions with the diode components. It is important to understand how a basic photodiode works in order to comprehend the operation of other photosensors used in vision systems. At this point a rudimentary understanding of semiconductor materials must be developed. The following paragraphs will detail some basic principles of semiconductor construction as they relate to the operation of the photodiode. This material may seem lengthy and over detailed, but understanding how photons of light can be absorbed by silicon and then subsequently converted to a measurable voltage output pro-

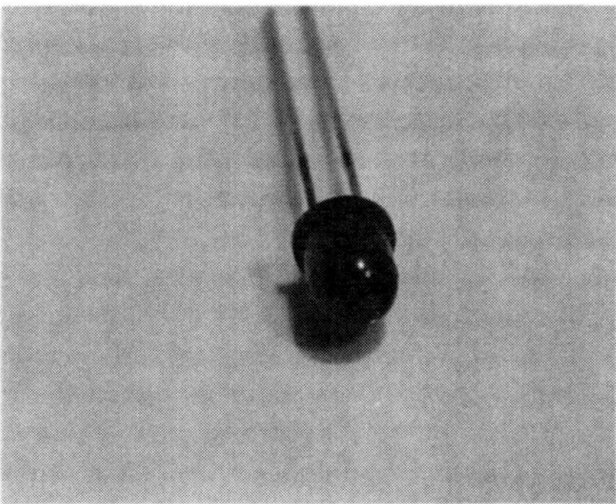

Figure 6.5. Common P-N photodiode used in machine vision systems.

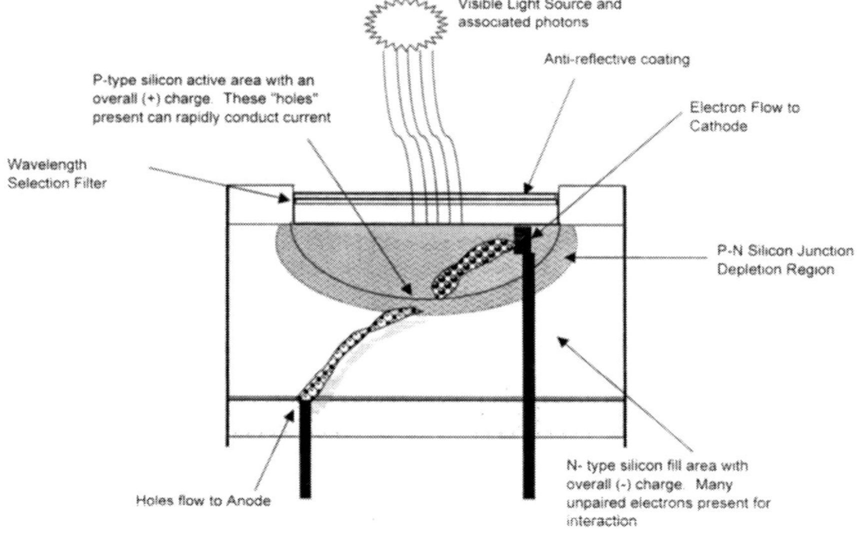

Figure 6.6. Cross-sectional representation of a typical P-N silicon diode and function of its components when struck by photons of light.

vides the reader with the necessary knowledge for conducting a critical review of commercially available vision systems.

Silicon atoms possess four electrons in their outer shell that can readily be involved in forming bonds with other elements. Compounds composed of nearly pure silicon are poor conductors of electricity due to the highly rigid crystalline structure of the silicon matrix. This matrix of silicon atoms keeps the electrons of their outer shell involved in covalent bonds with neighboring silicon atoms. A typical matrix of pure silicon is shown in Figure 6.7. In order to use silicon for the purpose of creating an integrated circuit, its composition and structure must be changed slightly so that it can become more conductive.

To achieve this result the silicon wafer materials used in the photodiode are doped with trace amounts of other elements to reach an overall positive or negative state. By imparting such characteristics into the silicon materials, they become conductive in nature, though not as much as a true metal, such as copper, would be. Because the newly treated silicon can conduct electricity, it is termed semiconductive. The N silicon of the photodiode must posses an overall negative charge structure to be functional. In order to manufacture the N silicon elements that are electron rich must

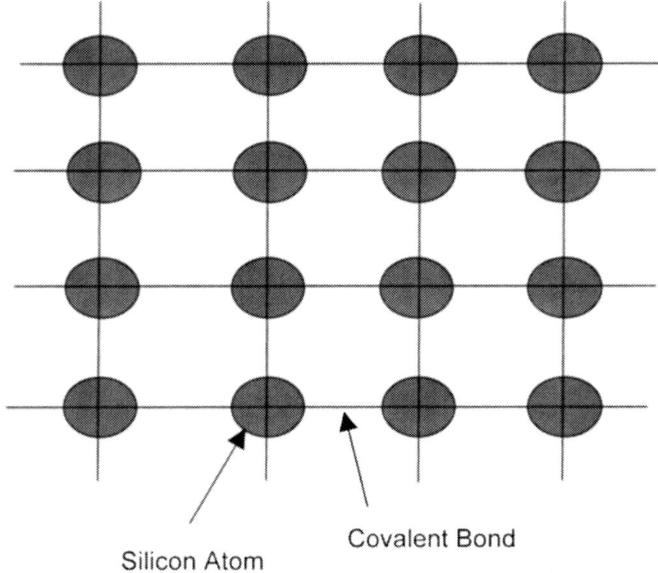

Figure 6.7. Two-dimensional representation of silicon lattice formed through covalent bonding in a pure silicon matrix.

be introduced into the silicon matrix. The element phosphorus, which has five electrons available to make bonds, is typically used to fill this need (see Figure 6.8). The extra electrons of the phosphorus that aren't involved with bonding to the silicon matrix can move throughout the structure and function as conductive electrons. For this reason the phosphorous is referred to as a donor impurity because it provides the conduction zone electrons.

Photodiodes also require a thin layer of P silicon to be present at the surface where light will be introduced. This area is called the active area of the photodiode. To manufacture P silicon that possesses an overall positive charge, a substance like boron is introduced into the silicon crystal matrix. Boron can only accommodate bonds with three other molecules. When introduced to the silicon its presence creates electron "holes" in the silicon-bonding matrix. Figure 6.9 displays a P-type silicon matrix and the hole effect created by the boron atom. Note that there is a free silicon electron present that could readily share its space with another electron. In fact, as the energy of the structure changes due to heat or other radiation, the holes will move around the matrix as electrons jump from other silicon atoms, or orbits, to fill the hole. Every electron movement leaves a

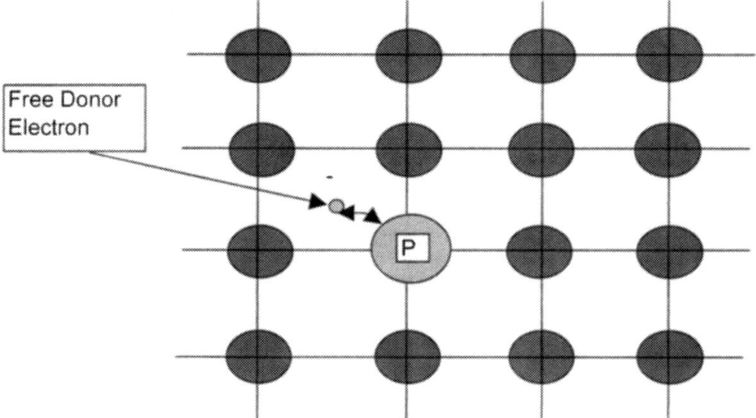

Figure 6.8. Two-dimensional representation of N silicon lattice formed through covalent bonding of phosphorus atoms in a pure silicon matrix.

vacancy which creates a new hole in a different spot in the matrix. Impurities such as boron are called acceptor purities because they make the structure ready to accept electrons (3). The hole formation process resembles a continuous session of musical chairs, except that in this case there is always a chair for the electron to sit in when the music stops.

When photons of light of sufficient energy (frequency) are absorbed in the active area of the photodiode, further electron-hole pair formation will occur. Photons of light that reach the silicon with sufficient energy cause the electrons of the inner shell to be moved to an elevated energy state, creating further hole vacancies. Because there is an electric field already present in the depletion region (where the N-P silicon materials interface), any particle that is liberated possessing a charge will be swept out of that area. This occurs due to the particle being attracted by the opposite charge region while simultaneously being repelled by the like charge minor carrier region of the matrix (3). When these positive- and negative-charged carriers flow in a "flip-flop" manner, there will be a measurable current produced. The measurable resultant voltage is proportional to the intensity of the light source because the number of electron holes that are created via photon interaction is dependent on the number of photons impacting the reactive layer of the silicon.

The voltage produced in the diode is an analog measure, and must be converted to a digital format for interpretation via a microprocessor. The

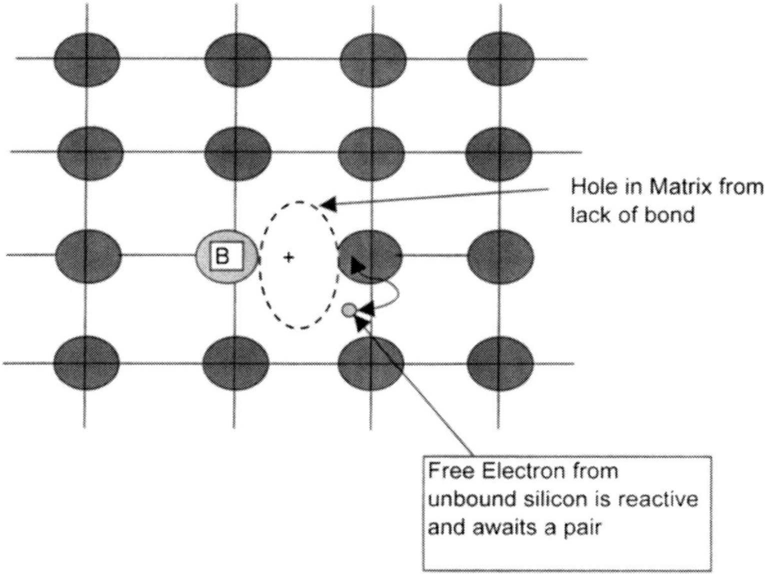

Figure 6.9. Two-dimensional representation of P silicon lattice formed through covalent bonding of boron atoms in a pure silicon matrix.

analog output from the gray scale images is usually digitized into 256 levels of increasing intensity (bits). The lowest level of the scale would represent a totally black image (no incident photons), whereas the highest level would be represented as a white image (saturation of the photodiode with incident photons).

One key concept to note at this point is that a base photodiode is essentially completely color-blind. Photodiodes may exhibit slightly increased responsiveness in certain areas of the visible spectrum, but this fact would in no way afford qualitative color detection. The photodiode is quantitative at this point and only accounts for the intensity of the light photons received upon its active surface, not the energy of the photon.

For a photodiode to sense color, the other reflected wavelengths of light that could potentially strike the photodiode and thus elicit a voltage response at the detector will need to be filtered out. Application of filters and other layers of reflective materials to the photodiode lens, and active surface layer of P silicon, allow for the sensor to be color-specific. Note that this specificity is produced via filtration of undesirable wavelengths of electromagnetic radiation, not due to some magic mathematical software trick. This is a concept that will ring true with all other sensing methods.

The typical filters available for photodiodes that are of interest in sorting out FM allow for the selection of infrared wavelengths, red wavelengths, green wavelengths, blue wavelengths, and ultraviolet wavelengths of varying intensities. The responsiveness of the photodiodes to each of these wavelengths may vary, but quantification of the number of photons striking the detector (voltage change) will allow for identification of the intensity of the light source.

Photodiodes can detect light photons present in a given area. Note that the diameter of a single photodiode is not very large (0.5 to 0.8 mm). This in essence is the sensing area available for use by a single photodiode. Obviously, to inspect a product stream, the sensing area must be larger than a few mm, so the photodiodes are arranged in linear arrays to cover a specific inspection area. Each photodiode present will be given an assigned position within the array that reflects its physical location and correlates to the voltage signal presented at the detector. Figure 6.10 illustrates a single strip of photodiodes arranged in an array, and an area detector array composed of several photodiodes. To prevent reflected light from affecting a neighboring photodiode, there are usually sheaths built around each photodiode. The sheaths rise above the lens of diode and work to shield the undesired reflected light. The sheaths coupled with the already large nature of the photodiode produce a decent amount of dead space in these

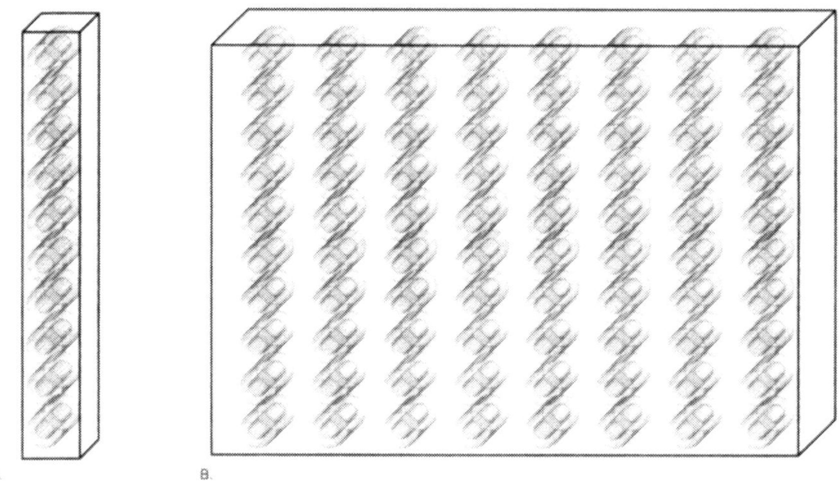

Figure 6.10. Linear arrays of photodiodes. A. represents an arrangement for a single line detector (1 × 10); B. represents an area sensor linear array (8 × 10).

arrangements. Using smaller diameter photodiodes is one way to compensate for this, and allows for more photodiodes to be packed into the same unit area. When more receptors (photodiodes) are present within an array of the same unit area, there will be a subsequent increase in the overall spatial resolution of the detector.

By interlacing the detector array with different types of individual photodiodes that are capable of detecting light intensities of differing wavelengths (colors) respectively, a bichromic or trichromic array can be made. Figure 6.11 illustrates a bichromic array of photodiodes capable of distinguishing intensities of more than one color of light present over a given unit area. The voltages can be digitized via use of an analog to digital converter for each photodiode output to compare the intensity of each of the wavelengths present. Software algorithms can then be used to identify any relevant patterns of light intensities or color distributions that may be present. The concepts that govern these software algorithms will be developed further toward the end of this chapter.

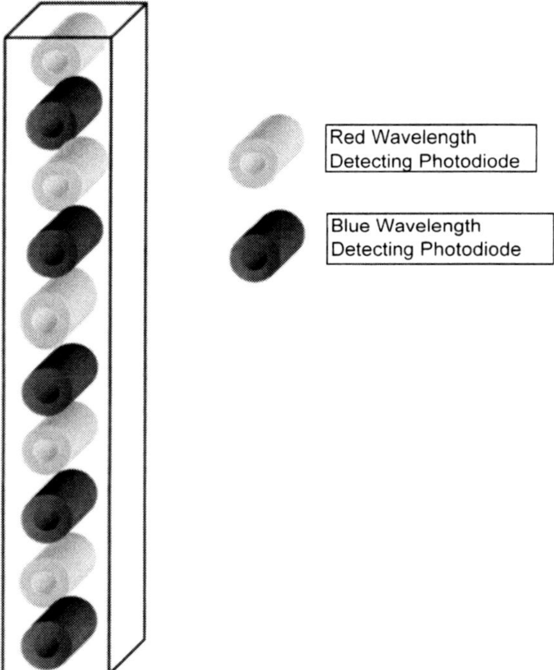

Figure 6.11. Bichromic photodiode array for identifying intensities of reflected red and blue wavelengths of interest.

The response time of a photodiode is extremely rapid, and it will have little bearing on the speed of inspection activities for foreign materials in food products. The typical rise-and-fall times for voltage changes in response to light photons striking a photodiode are on the order of 10 to 40 nanoseconds. Variations in this time are dependent upon the wavelength of light being presented to the detector. All in all, the speed and simplicity of the photodiode detector has allowed it to remain useful for machine vision applications over the past several decades.

6.2.4.2 Use of CCDs in machine vision systems

Charge coupled devices (CCDs) are specially fabricated semiconductor components that have been adopted by society for many uses. CCD cameras have become indispensable components for sensing wavelengths of light in a rapid manner, and they provide a readily digitizable signal format. The components of CCD cameras are similar to those of a film-based camera except that the film is replaced by a silicon wafer containing many individual photosites within pixels. The typical construction of a CCD photosite is depicted in Figure 6.12. The photosites for a CCD are composed of layers of N and P silicon and respond to differing light intensi-

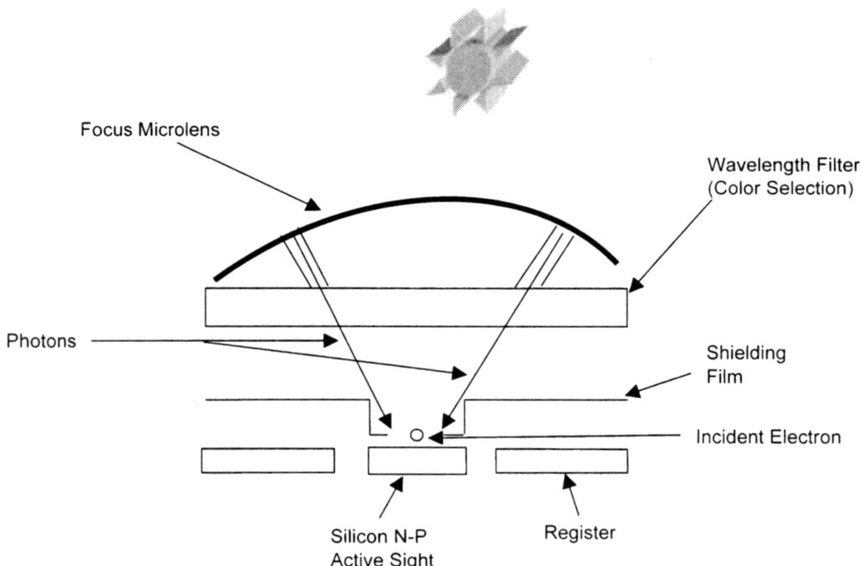

Figure 6.12. Typical components and function of a photosite in a CCD.

ties in a manner similar to that of the photodiode; that is, via production of electron hole pairs. The main difference in a CCD setup, however, is that the pixel has the ability to store (bin) electrons that have been liberated into the conducting zone.

The free electrons produced will stay in what is known as a potential well until the CCD is electronically shuttered (discharged) during the readout phase. The CCD performs the readout function by inducing an electrical current across the array that will shift rows of image information in a parallel fashion, one row at a time, to the serial shift register. The serial register then utilizes a second clocked electrical current to sequentially shift each row of image information to an output amplifier as a pixel-by-pixel serial data stream. This measured voltage will need to be passed through an analog-to-digital converter, which creates a digital electronic representation of the incident photons. This process is graphically represented in Figure 6.13.

The photosites of a CCD are much smaller than their photodiode counterparts. The photosites typically measure <10 micrometers in diameter in modern CCDs. This affords a greater bunching of pixels per unit area when compared to linear arrays of photodiodes. The CCD pixel arrangement provides for far greater spatial resolution and far less dead areas within the image when compared to a photodiode array. The already small size of the pixels will continue to decrease as technology evolves, and this will allow for CCD cameras possessing incredible resolution factors.

The CCD configurations available can be formed into detection arrays much like photodiodes. A CCD known as a line-scan camera is simply a single long strip of pixels spread out over a large distance. This design is quite suitable for capturing data from a process in motion (positioning on a high-speed conveyor). Other CCD cameras (known as area detectors) of various configurations are commercially available. Current models range anywhere from 455 × 512 pixel configurations to 2048 × 2048 pixel configurations. These systems offer incredible spatial resolution and outstanding overall performance for image capture.

As previously noted, the CCD requires filters for selection of the desired wavelengths of visible light. Most commercially available CCDs accomplish this task via grouping of pixels in mosaics of red-, green-, and blue- (RGB) sensitive spatial arrangements. The output from RGBs in an area is usually combined to represent a unit area in color. The typical mosaic is depicted in Figures 6.14 and 6.15, and has provided for excellent differentiation of the color components of observed images. Usually the analog output for each specific color intensity sensed via the pixels is

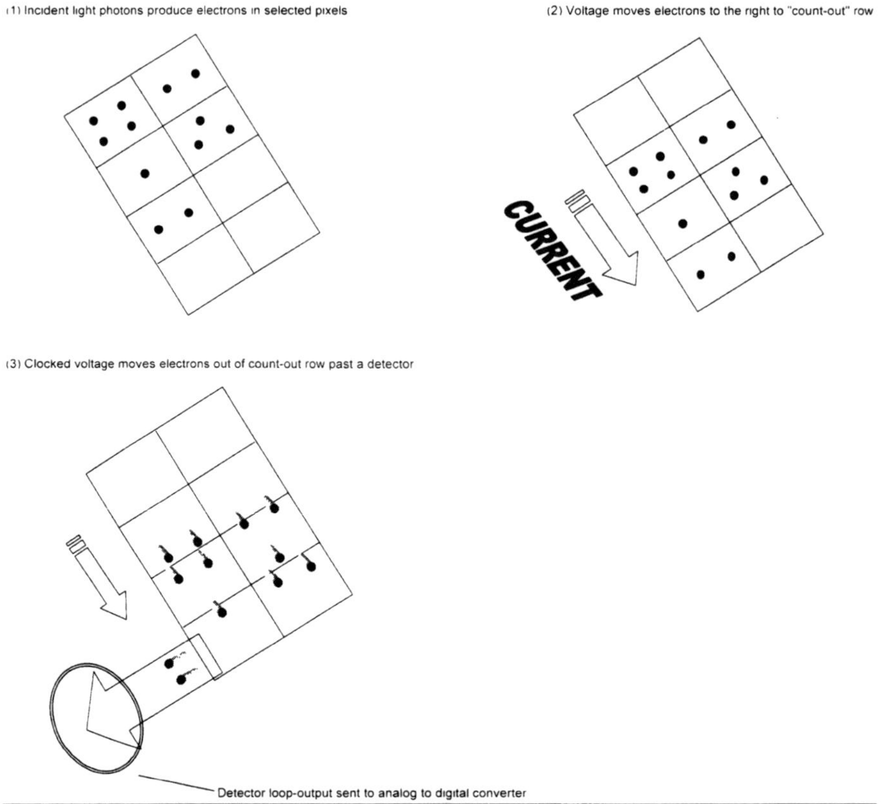

(1) Incident light photons produce electrons in selected pixels

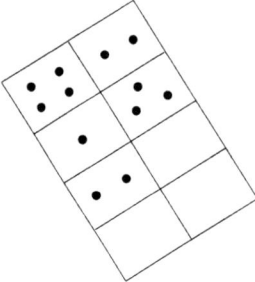

(2) Voltage moves electrons to the right to "count-out" row

CURRENT

(3) Clocked voltage moves electrons out of count-out row past a detector

Detector loop-output sent to analog to digital converter

Figure 6.13. Binning, register, and readout of electrons stored on a CCD.

Mosaic Capture

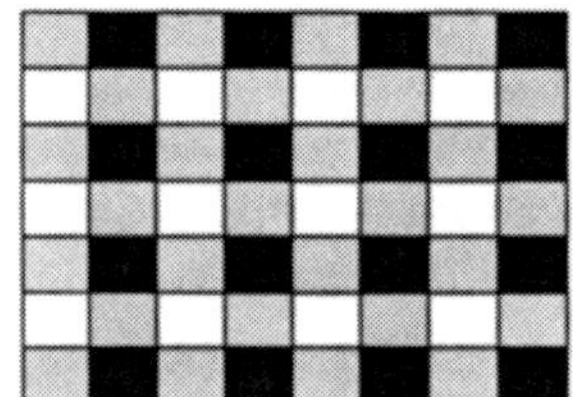

In conventional systems, color filters are applied to a single layer of photo-detectors in a tilted mosaic pattern.

Figure 6.14. Illustration of mosaic capture in a CCD.

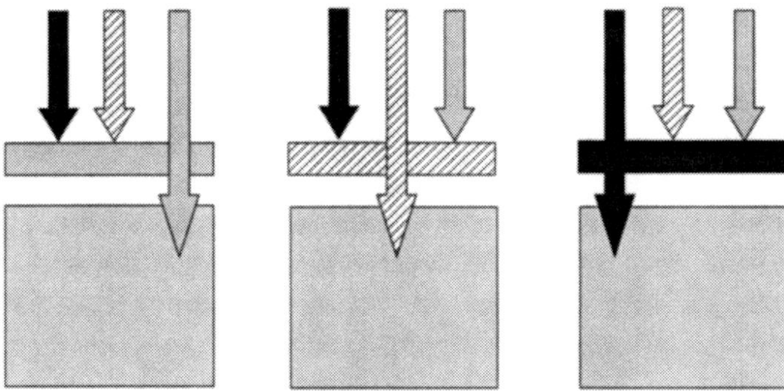

The filters let only one wavelength of light - red, green or blue - pass through to any given pixel, allowing it to record only the color wavelength not attenuated.

Figure 6.15. Light filtration in the mosaic of a CCD.

digitized into eight or more corresponding levels (bits) of intensity. This is done at the digital-to-analog converter, and is based on a predefined classification scheme. It is possible to differentiate further classes for each color, say up to 64 bits per color, but the overall increase in complexity by doing so usually would not result in better separation of foreign materials sorted via color.

6.2.4.3 Use of CMOS systems

The CCD has undoubtedly been the imaging system of choice for industrial vision systems for the past two decades, but advances in the manufacture of complementary metal oxide semiconductor (CMOS) image sensors have proven to be tantalizing. A CMOS imager works in a very similar manner to a CCD camera, in that the initial reception of the light photons and conversion to electrons occur in the conduction zone of an N,P silicon matrix. The major advantage that CMOS image sensors enjoy over their CCD counterparts is that they are assembled with the same technology and equipment used to assemble microprocessors. By imbedding multiple circuit paths in the chip, the CMOS has the ability to integrate a number of processing and control functions, other than collecting photons, directly onto the sensor's integrated circuitry. The beauty of this is that there is a great deal of functionality delivered by the CMOS sensor at a fraction of the cost of a CCD.

The CMOS designs that are currently being manufactured use active pixel sensor (APS) technology. In this arrangement both the photodiode and readout amplifier are incorporated into the pixel. This enables the electron charge accumulated via the photodiode to be converted to a voltage at the pixel sight. The voltage will be amplified at the pixel and then transferred via a more traditional wire circuit to the analog signal processor on the chip. All of these added features do come at a cost though. Each pixel now contains a photodiode and multiple transistors organized into a checkerboard fashion. These components, coupled with the other circuitry, make the CMOS design bulky and effectively dilutes the active sensing area. This design does, however, enable signals from each pixel in the array to be read directly via x,y addressing techniques, which is not possible with current CCD technology.

6.3 Frame grabbers and other external equipment such as photoeyes and signals from PLCs

The frame grabber as defined by the Automated Imaging Association (AIA) is "a device interfaced with a camera storing in memory, on command, sampled video converted to digital signals." Essentially, the frame grabber is the part of the system that takes the camera information and turns it into something that is readily comprehended by the computer. The typical components of a frame grabber are: an interface section for the camera, on-board memory for storage of the information from the camera, and a host computer interface. Some of the more advanced frame grabbers available have provisions for onboard signal conditioning and other digital signal processing capabilities via the addition of microprocessors. The frame grabber components are typically housed on a computer board and can be mounted either internally or externally to the host computer.

The function of the frame grabber is usually triggered by either an internal timing event or input from an encoder or photoeye measuring the presence of the material in the conveying system. Once the frame grabber is triggered, it will capture and condition the analog signals from the camera, or cameras, and use the appropriate system bus standard to upload an encoded digital image to the host computer. Frame grabbers by themselves can impart a certain amount of noise into the vision system if they are not properly synchronized with the camera or timing signals. Noise can also be introduced by subjecting the frame grabber to elevated temperatures or other harsh environmental conditions. Recognize that the cable connections between the camera and the frame grabber, as well as from the triggering device to the frame grabber, can become a spot which can cause poor performance. Corrosion of contacts, distortion of fiberoptics through poorly routed cables, and loose connections are typical problem areas. Other frame grabber watchouts deal with incompatibilities between frame grabber boards, cameras, and PCs. This can be especially problematic if in-house resources are attempting to build a custom machine vision system on a shoestring budget in an attempt to address a specific issue. Usually it is best to source the camera and the frame grabber as a package deal so that unexpected issues do not crop up. Alternative component sourcing strategies usually result in finger pointing between vendors, and the processor being stuck with a poorly functioning vision system.

6.4 Machine vision computers, image processing software, and reject outputs

The early monochromatic machine vision systems in operation at the beginning of the PC era ran effectively on 8086-type computer processors. These processors were relatively slow in calculating matrices, and could not handle large amounts of information. The capability of such systems is blown away by modern technological advances in microprocessor technology. Moore's law of increasing complexity of technology has remained valid, for now computer processors with clock speeds of >3 gigahertz are commonplace in today's systems. The complexity of technology continues to increase, and there has yet to be a sustained slow period in product improvements. The architecture of the board and system bus of some modern computers allows for external inputs of more than one gigabyte of data per second from frame grabbers. To further leverage this technology, motherboard manufacturers continue to design amazing multiple processor-based boards capable of incredible data segmentation and analysis speed and accuracy.

6.4.1 Software components of machine vision systems

The software platforms and algorithms that are available for modern machine vision systems have evolved substantially as well, but are somewhat lagging behind compared to the hardware components. Realize that the software is the critical component of the machine vision system that enables the automated product analysis. The software is also the initial source of the rejection output signal as well. Poorly written software algorithms can result not only in missed rejections of defective products, but also in expensive false rejections and subsequent lack of operator confidence in the vision system. The software algorithm components of the machine vision system will, by far, represent the greatest expenditures of money and time when developing a new system. The potential of today's machine vision systems essentially is not limited by processor potential, but more by the programmer's creative delineation of what represents acceptable and unacceptable characteristics. This is especially true in situations where the products to be inspected and the potential defects to be identified are inherently variable in nature. Many of the imaging techniques used for discriminating good product from FM will be discussed later in the chapter, but in the meantime realize that sometimes the most complex software algorithms developed for an application may not be the most appropriate solution for the problem (2).

Simple and stable are also the credo for choosing the operating system (OS) that the computer and imaging software will work within. The latest and greatest multitasking OS is usually not necessary for the computer component of the vision system to operate well. In fact the latest and greatest OSs have a tendency to operate no faster than previous versions because of added new features and ready networking components that are rarely used on computers dedicated solely for machine vision systems. The newer OSs usually use more random access memory (RAM) as well, and this is a precious commodity for a machine vision host computer. Any random RAM corruption or other prioritization conflicts presented by the OS could potentially cause isolated issues with pixel loss or output distortion.

6.4.2 Ensuring the computer is dedicated for the machine vision system

The host computer should also be dedicated to the task at hand—sorting critical components and defective units from a high-speed production line. Other than downloading reports or allowing vendors to debug program glitches remotely via modem, the computer should be devoid of any other programs or tasks (i.e., spreadsheets, e-mail, web browsers, games, etc.). A stern administrative expectation should be set stating that the computer terminal is not intended for general use, but rather for use by authorized personnel only. As a general rule any person caught playing games or perusing the Internet on the machine vision host computer should be dealt with in a serious manner. My personal favorite way of dealing with this issue is to require the offender to sit alongside the camera and perform mundane 100 percent visual inspection activities for a few hours to help "validate" the system's reject function. Usually the offender will be forever reformed, and will have gained a newfound appreciation for the vision system's components.

6.5 Reject mechanisms

The best machine vision system is only as effective as the reject mechanism that executes the segregation. Modern high-volume machine vision systems typically use air-actuated pusher rams, vacuum paddles/bars, or expansive networks of electronically actuated solenoid air valves that act as reject mechanisms. The reject tracking and timing is critical for successful operation of the machine vision system. If the system is not accurate at rejecting potential contaminants, there is obviously potential for the proliferation of more serious issues. Conversely, if the reject system is accurate but offers little precision, serious cost implications can result from the scrapping of perfectly good product. I once had the misfortune to witness a two-day-long production period where 35,000 Kg of costly imported agricultural products were run over a poorly functioning color sorter. The algorithm was equipped with poor thresholding logic, and the air rejectors were being fed from an unsuitably supplied air source. The end result of the operation was ten hours worth of finished product held for undesirable seed inclusions, and a net six percent yield loss in raw materials due to false rejects.

The reject mechanisms of any machine vision system are going to require routine inspection to ensure that they are functional. This can be quite challenging on certain systems due to the vast array of rejectors that may be present (some systems have as many as 256 individual air ejectors). Consider monitoring the supplied volume and quality of compressed air that feeds the system as well as checking any tubing connections for leakage or breakage. Coupling these inspection activities with systematic valve function tests over time, the reject mechanism should remain functional. Ideally the machine vision system would be interlocked in a manner that, if air pressure dropped below critical limits that assure proper valve response times, the sorting would stop and an alarm would result.

6.6 Typical sorting scenarios utilized by machine vision systems in identifying FM

Just as was the case in the physical separation chapter (Chapter 3), the key to successfully applying a machine vision strategy to isolate good product from FM lies in knowing what the repeatable differences are between the two. This difference must offer a high degree of contrast from the background image as well. The differences also need to be readily imaged via the sensor(s) utilized. This is important in order to provide a measurable response that the host computer can apply various mathematical operators to for signal interpretation. Typical differences exploited by savvy machine vision system integrators to separate FM from materials are:

1. Differences in color
2. Differences in shape
3. Differences in size
4. Differences in absorbance or transmittance of nonvisible electromagnetic radiation such as IR or UV
5. Pattern recognition of surface features, color distribution, or nonrandom shapes

Sometimes it may be necessary to combine several subtle differences in each classifier category to separate an FM from a food matrix. Algorithms of this nature are highly complex, and usually take a long period of time to properly tune in, even when a full host of good and bad samples are at the technician's disposal. Before diving into complex algorithms, however, it is important to understand how a simple algorithm would operate in differentiating FM from a product. The case study below will present such a case, then work through the initial development of more complex algorithms.

6.6.1 256 gray scale thresholding example

For this example a common, relatively simple technique known as thresholding will be applied using a gray scale (256) image. The goal is to identify paper stickers stuck on the surface of purchased fruits that are slated for use by a food processor. It is common when buying certain grades of fruit products on the open market to occasionally encounter an applied price look-up sticker (PLU) on the surface of the fruit. For some product types these can be a quick source of consumer complaints if they are not sorted out or removed prior to processing.

Figure 6.16 displays the defective fruit product. Note the size of the sticker and its placement. For this example it will be assumed that the stickered side of the produce will always be in the view of the camera at some orientation (this will keep the topic more comprehensible than introducing a 3D machine vision tactic at this point). A well-trained human inspector could reliably spot this defect if the fruit was presented to the individual in a well-lit area. This fact makes using a machine vision system a potential good fit for continuous inspection of the product stream.

A monochromatic line-scan CCD machine vision system will be used for this inspection. A direct front lighting technique has been chosen to better illuminate the surface of the fruit. The system is turned on and the camera is allowed to capture a gray scale shot of the background. The digital signal that represents the background is displayed in Figure 6.17. Note

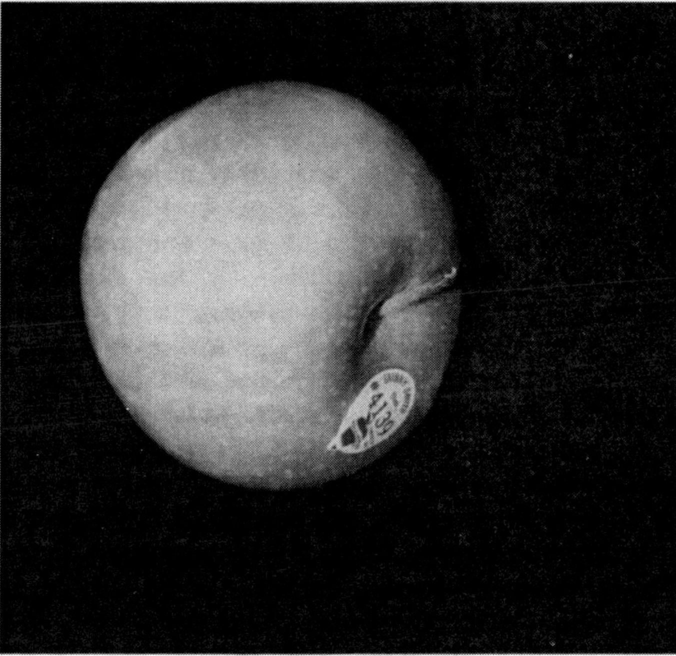

Figure 6.16. Defective produce with a PLU sticker.

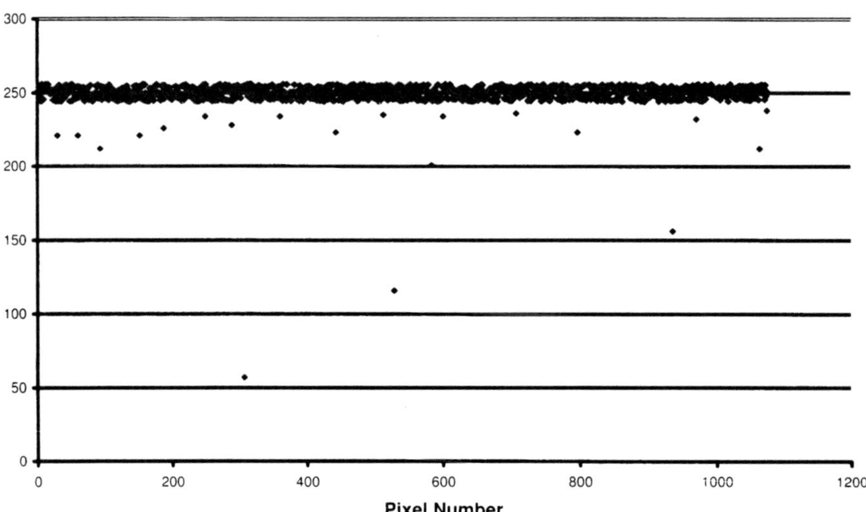

Figure 6.17. Running 256 gray image of the vision system's background (conveyor).

that this background does not utilize a great deal of the 256 gray scale range. The light source is now adjusted, and the background is nearly "zeroed out" to all but some minor "noise" present in the image.

Satisfied that the system is performing properly in acquiring an image, "good" fruit is visualized and a gray scale baseline is developed by the system. This process is similar to training a new human inspector in how to separate a good product from a defective one. Unfortunately, it is more difficult in this case because the machine vision system has to learn quantitatively how to ignore the background, and then identify the rate of sampling necessary to compensate for the product's motion. A large variety of reference samples are analyzed in this stage, and the observed distribution of gray scale intensities are directly measured and recorded by the system. The actual observed high and low intensities, in theory, could be used to develop a maximum and minimum intensity value for the product. Usually it is more appropriate, however, to utilize a statistical technique to account for variation in the product when developing the product image specification. Calculating the sample mean and variance of the pixels captured is usually a good starting point for defining the product signal. A histogram of the product signal produced during the training period is displayed in Figure 6.18.

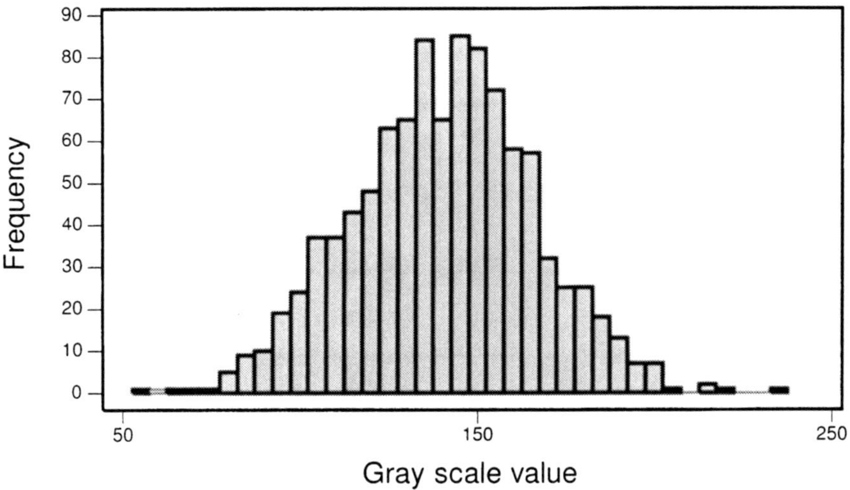

Figure 6.18. Histogram of the good apples that were discovered while analyzing hand-selected samples.

The defect that is to be sorted out will also need to be characterized in a similar manner. The defective fruit was taken from the sorting bin and analyzed by the system. The resultant signal for the produce (minus the background) is displayed in Figure 6.19. Notice the abrupt change and signal difference that the system identifies. This correlates to the presence of the sticker quite nicely, and should afford differentiation by the machine vision system. For further diligence the stickers, now peeled from the fruit, were placed on a background and presented to the vision system. The smallest white portion of the sticker in any dimension was found to be eight pixels in length, and random presentation of stickered fruit routinely afforded line scans with 20 white pixel run segments. A reverse color scheme will be applied for better imaging of this area of interest. The line scan for the reverse color image is displayed in Figure 6.20.

Now that all the relevant data about the nature of the produce and the defect has been collected, some key facts can be assembled to define how the system will identify the defects. Based on the statistical analysis of the product pixel distributions versus the FM, a simple thresholding strategy can be used to identify defective pixels. This will be done by writing a software algorithm that will assemble all of the pixels from a scan, compare them to a value in a look-up table, and flag any pixel that possesses

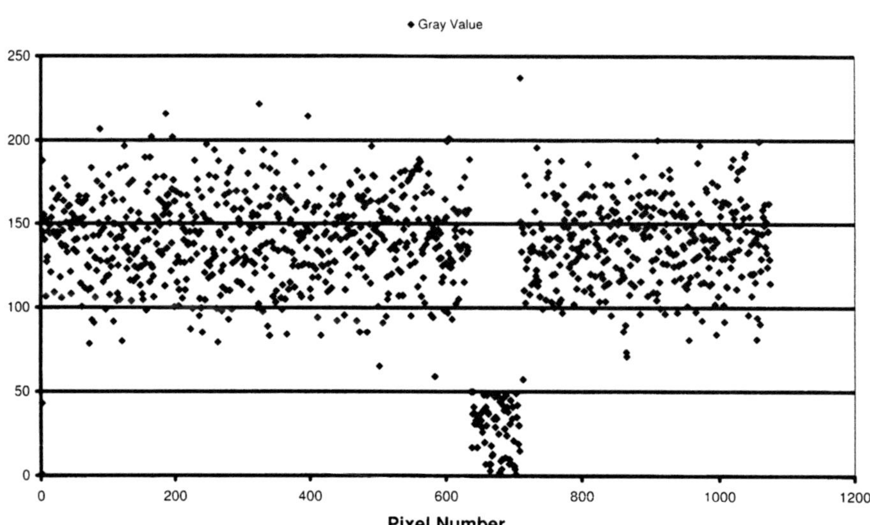

Figure 6.19. Pixel values from a line scan of the defective apple. Note the intense drop in gray scale counts in the area where the sticker is scanned.

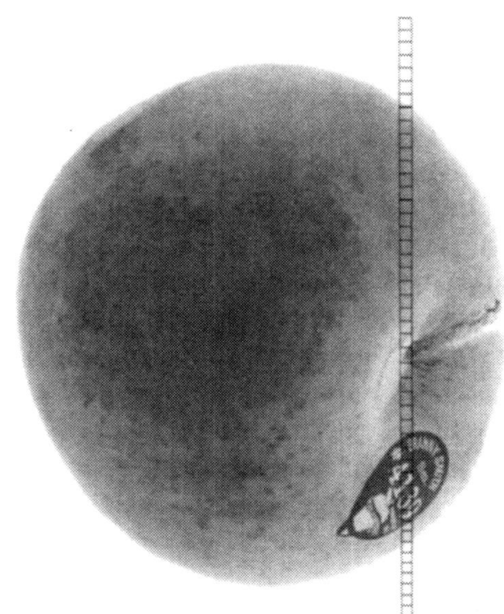

Figure 6.20. Representation of the line scan pixels captured for making Figure 6.19.

an intensity value between one and 49 as a defective. Results for the defective and non-defective pixels will be coded in binomial format where a value of zero represents good and one represents the defective condition. This operation is represented in Table 6.1.

The algorithm then specifies that each column of transformed line scan data be counted. If there are greater than 20 pixels present in any single line scan with a value of one, the algorithm flags the logic for the rejection signal. This is obviously a very simple algorithm, but it is still used by many manufacturers of machine vision systems and food processors alike, to identify FM. More advanced adaptations of thresholding applications allow for on-the-fly recalculation of the product baseline and analysis by color. Realize, however, that this type of algorithm is not identifying the shape, contour, or variation components of the image. Vision systems

Table 6.1 Example of a basic thresholding algorithm used to isolate pixels that represent defective components.

Pixel	*Raw Gray Scale Data*	*Logical Operator*	*Program Value*
1	131*	FALSE	0
2	158	FALSE	0
3	25**	TRUE	1
4	105	FALSE	0
5	128	FALSE	0
6	137	FALSE	0
7	40	TRUE	1
8	147	FALSE	0
9	118	FALSE	0
10	158	FALSE	0

Sum of TRUE Program Values for area = 2

Logic applied for pixel classification:
* Individual Pixel >49 = "FALSE" returns Program Value = 0
** Individual Pixel < 49 = "TRUE" returns Program Value = 1
A Gray Scale value of <49, in this example, represents a defective unit to which the program must assign a value.

Logic applied for the disposition of an area based on the sum of the pixel classification operations:
Scenario 1. Sum TRUE < 3 = "ACCEPT" No Reject IO triggered
OR
Scenario 2. Sum TRUE >3 = "REJECT" Reject IO triggered

The disposition in this case would be to "ACCEPT" the area and no reject signal would be sent to the process.

based on thresholding algorithms simply employ a counting function that bins good and bad pixels together in a scan and makes a reject determination. This is important to understand because, when using a similar thresholding scheme, it is imperative that the product, the defect, and the background not have undue overlap. As a result, situations involving elevated amounts of overlap usually have a greater number of false rejects.

6.6.2 RGB imaging and thresholding

As stated previously, a color machine vision system can be used for thresholding operations as well. This scheme will require a much higher complexity of algorithm to accomplish the desired task, partly because each color output (RGB) from the product signal will need to be dealt with by the algorithm differently. A visual representation of what a color thresholding system would look like for segmentation of nectarines with product stickers is shown in Figure 6.21. The RGB color cube is used to display the results as they are segmented into their respective color bins. It is important to note that in order to represent the subtractive colors of the spectrum (CYM) from the RGB input, an additional layer of mathematics will be necessary. The mathematical formula for deriving per-pixel intensities for each subtractive color will add further complexity and com-

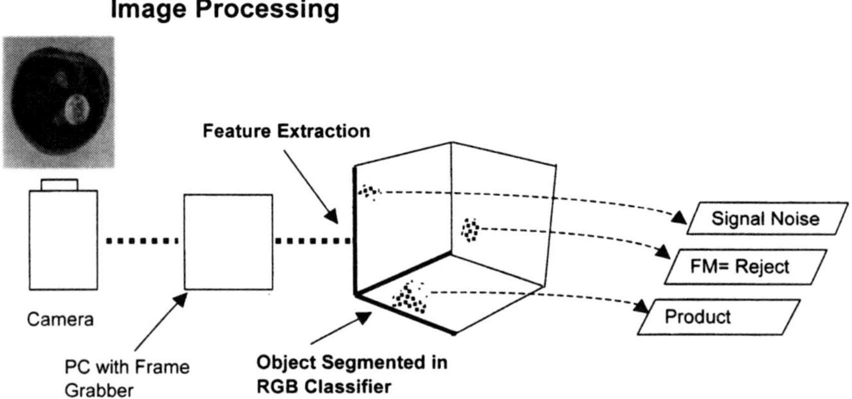

Figure 6.21. Conceptual presentation of defect sorting via color for PLU sticker identification.

putation time to the algorithm, but this may be necessary in order to differentiate products from FM.

6.6.3 Beyond thresholding—compiling pixels into objects

As stated earlier, thresholding operations only provide output that summarizes pixels based on their ability to meet or not meet a predefined numerical classification. The resulting file is still pixel information, and does not yet offer any reference to the presence of an object. In order to make the machine vision system understand if these pixels are randomly dispersed or part of an object, a grouping methodology must be used. These steps are typically referred to as segmentation of the image and feature extraction (2).

The segmentation is usually performed with the output from a thresholding algorithm. When the thresholding output is received, algorithmic operators will be utilized to check for pixel groupings known as blobs. These are usually based on a nearest-neighbor-type connectivity scheme (Figure 6.22) which will analyze adjacent pixels of a reference pixel for similar characteristics (2, 8). This type of scheme is desirable for spotting true patterns within the sea of pixels versus random noise within the image. As stated previously, the output from a machine vision system will possess random noise and other factors that can affect the thresholding results, providing potential false positive pixels. Once the blobs have been identified they, too, can be further segmented, if necessary, using an algorithm similar to the nearest-neighbor algorithm presented above.

The resulting blobs are then characterized by their size, continuity, or other measurable operator to provide ready information about the image in a vector format. The process of extracting the useful image components of the blobs is known as feature extraction (2), and will provide the data necessary for classification of the object visualized. The final classification of the object will be performed via comparison of the object features with those contained in a mathematical model. The object will be classified into a group based on its degree of correlation with this model. The output from this grouping procedure will usually trigger a counting function, or the system's reject mechanism, the process summarized in Figure 6.23. This type of analysis is essentially the basis for all shape and morphology recognition schemes. Obviously the concepts presented here are quite

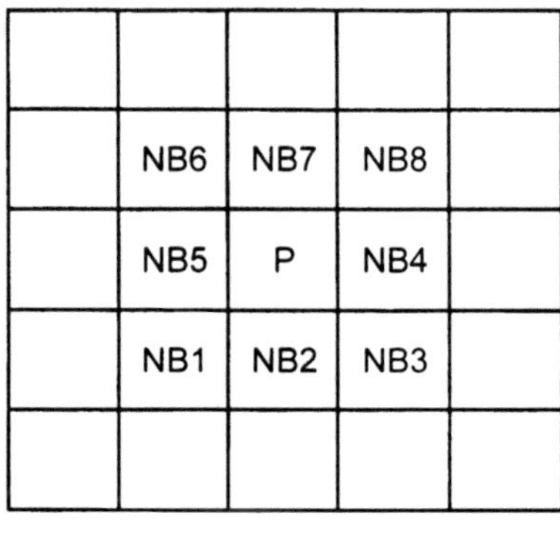

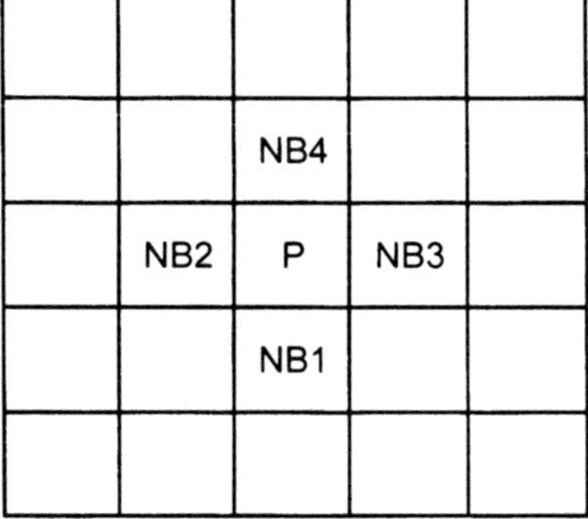

Nearest neighbor comparison for segmentation via connectivity

Figure 6.22. Example of understanding if pixels are actually grouped together as part of the same object, or randomly distributed. The nearest neighbor technique is used in this example.

simplified, but they should provide for comprehension of how the systems work to classify objects in the field of view.

6.7 Current machine vision applications to identify FM in foods

There are numerous commercial systems available to sort defects and foreign materials from acceptable food products. The turnkey applications range from segmentation of rocks from produce such as peas and blue-

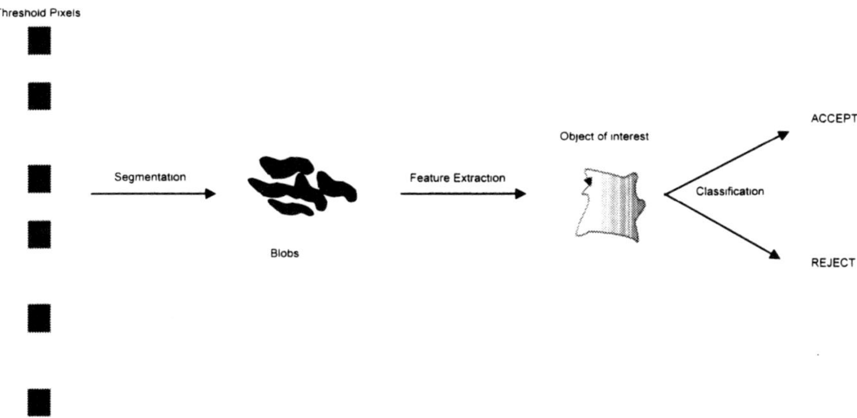

Figure 6.23. Illustration of how images can be segregated, have their features extracted, and classified via software operators.

berries to sorting sticks, stones, and feces from coffee beans. Many of these systems are costly, but they offer far more capability than merely sorting foreign materials from food products. The added functionality offered by multitasking machine vision systems can be the tipping point when it comes to cost justification of the equipment. Multitasking machine vision systems are being used to ensure label position and correctness, verify fill level, verify counts, and check other quality attributes such as texture and continuity.

There are far more applications that have been developed by the internal resources of major corporations that are proprietary in nature, and not publicized. The flexibility of the machine vision technique allows for the quick development of novel sorting techniques by organizations. This process is accelerated by the vast internal knowledge, and sometimes large defect libraries, that the corporations possess related to the characteristics of their products. These techniques will usually remain proprietary for long periods of time because they usually help deliver a sustainable advantage over a competitor.

6.8 Identification of FM in packaging materials

The other area where machine vision systems have been successfully applied is in the monitoring of foreign material inclusions in packaging material streams. Optical scanning equipment has been successfully utilized for decades to identify glass fragments, liquid contaminants, and other foreign materials from food packaging. Application of this technology has assuredly resulted in a mass decrease in consumer glass finds for companies that produce beverages and foods packaged in glass. Vision technologies have also been successfully used to inspect caps and lids for liner inclusions, voids, and stringing at both the manufacturer and processor, though for some reason this application is not widely publicized.

Can manufacturers have used machine vision systems for years to check the distribution of the coatings present at the product contact surfaces. These systems have also been effective at identifying nonrelated FM inclusions, as well as metallic FM inclusions, from the can production process. Several systems are commercially available to perform this task, and they can be seamlessly incorporated into filling lines.

A third packaging component that can be readily inspected with use of a vision system is plastic films and other multilayer webs. A simple line scan camera can be placed in position to view the film as it is being coiled or uncoiled. Placement at this point can flag the presence of FM as well as other defects prior to the defective film being threaded into the packaging machine.

Works Cited and Select Resources

1. Balban M, Yeralan S, Bergmann W, Otwell S. 1994. *Automation of Shrimp Quality Evaluation.* In: Developments in Food Engineering. New York, N.Y.: Blackie Academic & Professional. 1,085 p.

2. Chan JP, Batchelor BG. 1993. *Machine Vision for the Food Industry.* In: Food Process Monitoring Systems. London, UK: Blackie Academic & Professional. 220 p.

3. Chang R. 1994. *Chemistry. 5th ed.* Arlington, Tex.: McGraw-Hill, Inc. 1,088 p.

4. Hoshino M. 1994. *Wireless Image Transit Unit and High Speed Inspection System for Inner Side of Container.* In: Developments in Food Engineering. New York, N.Y.: Blackie Academic & Professional. 1,085 p.

5. Huang Y, Whittaker AD, Lacey RE. 2001. *Automation for Food Engineering.* Boca Raton, Fla.: CRC Press. 240 p.

6. Key Technology, Inc. 2001. *Field Reports: When the Beans are Green, Sorters are Keen.* Food Engineering (3):online. Available from: <www. food engineeringmag. com>.

7. McConnell RK. 1999. *Lessons Learned from Five Years of Successful Color Machine Vision Inspection of Automotive Fuse Blocks.* Arlington, Mass.: Color Machine Vision. 12 p.

8. Majumdar S, Luo X, Jayas DS. 1997. *Image Processing and Its Applications in Food Process Control.* In: Computerized Control Systems in the Food Industry. New York, N.Y.: Marcel Dekker, Inc. 597 p.

9. Park B, Chen Y. 2000. *Real-time Dual-wavelength Image Processing for Poultry Safety Inspection.* J of Food Process Engineering (11):329–352.

10. Vander AJ,, Sherman JH, Luciano DS. 1990. *Human Physiology. 5th ed.* Arlington, Tex.: McGraw-Hill, Inc. 832 p.

11. Wedd M. 1993. *Particle Sizing: Machine Vision for the Food Industry.* In: Food Process Monitoring Systems. London, UK: Blackie Academic & Professional. 220 p.

CHAPTER 7

X-ray Examination of Foods for Foreign Materials

The application of x-ray technology in the food industry boomed in the last half of the 20th century. X-ray applications are now becoming common practice in all segments of the food industry, and their prevalence continues to increase. The chapter is dedicated to providing the reader with a thorough discussion of the current technology, so applications related to foreign material control will be conceptually understood. Hopefully provision of a conceptual basis to the subject will afford the reader thorough knowledge prior to pursuing a system installation. The chapter essentially consists of four sections, organized in a build manner to better facilitate conceptual understanding of x-ray technologies. The four areas covered within this chapter are:

1. History of x-ray applications—conceptual development of the underlying principles
2. Overview of x-ray systems utilized in the food industry
3. Principles of automated imaging and inspection of x-ray results
4. Important considerations when choosing a system

It is recommended that the reader first touch base with the concepts and equipment discussed in the machine vision chapter (Chapter 6) so that the principles governing image capture, threshold analysis, and morphology are more tangible.

7.1 Introduction to x-ray inspection techniques and their use

X-ray inspection is currently one of the pinnacle technologies for detecting imbedded contaminants within food products. There are several benefits to using an x-ray system versus a more traditional metal detection arrangement. X-rays are far more reliable at culling out stainless steel and nonferrous contaminants from food products than traditional metal detection strategies. This is especially true for applications where detection of contaminants such as metal wire and swarf is being considered in a conductive food product.

The reason for choosing an x-ray system for this and other applications will become apparent, but for now it is important to understand that x-ray system sensitivities are based on identifying density gradients within the product. X-rays are not affected by the conductivity or magnetic nature of the contaminant (as in the case of metal detectors). Because of this density-dependent relationship, for which we will provide a more thorough discussion in later sections, x-rays can also readily detect glass fragments, stones, and bones that were once invisible when imbedded in solid products or contained in products with a turbid nature. X-rays can even identify fractional contaminants such as rubber, hard plastics, woody materials, and certain packaging-related contaminants in cases where the density gradient of the product permits. These types of contaminants are elusive to almost all other commercially available technologies available for use, with the exception of extremely low throughput high cost magnetic resonance and ultrasonic inspection systems that are still in development.

The potential range of x-ray applications must also be a consideration when evaluating the use of the technology. X-ray systems can be used for the inspection of finished product within its final packaging configuration. The presence of FM within metal cans, foil packages, glass containers with metal lids, and product pipelines are all feasible when using x-ray technology. Newer technologies offer the opportunity for inspection of full cases of product. The advantages and disadvantages of where x-ray technology is to be implemented will be discussed in later sections, but having the flexibility to inspect product at any point along the process line is an outstanding benefit.

The largest barrier for adaptation of x-ray inspection technology by a processor is the cost of the unit. The recent insurgence of the adoption of x-ray systems by food processors is rooted in the continually lowering cost barrier of the technology. The cost of x-ray systems, possessing excellent reliability and adaptability, are beginning to reach parity with those of high-end metal detectors. This trend is largely due to the continuing maturation and evolution of the semiconductor (SC) and integrated circuit (IC) manufacturing industries. Essentially the food industry is reaping the benefit of the fact that development costs for the basic electronic technologies that are integral components of x-ray systems has been recaptured many times over by the IC and SC manufacturers. As the manufacturers' economy of scale improves, further cost reduction and improved quality can be expected and, as advancements in the SC and IC industries continue, namely in data storage, retrieval, and advanced processing, the cost of components and overall capability of x-ray systems available to food processors and distributors will continue to improve as well.

The latest x-ray machines aid in the justification of their higher initial cost by offering expanded functionality beyond traditional FM control. X-ray inspection routines are now being used in place of check-weigher applications, off-line package integrity check stations, and evaluation of other quality attributes. Current systems have been coupled with advanced software to perform real time 100 percent inspection of many quality attributes. The current extended functionality of x-ray systems allows for inspection routines such as: (1) enumeration of objects in sealed packages, and (2) screening for mislabeling situations (identifying that a different product is present through density characteristics). By having the x-ray system perform such routines in addition to monitoring for product contamination, the displacement of labor is usually achieved, resulting in tangible cost savings for the organization. These savings can be used to help justify the equipment expense. It is possible, as reliable x-ray technologies become more prevalent in the food processing industry performing these functions, that the baseline value entitlement relationship of consumers and customers will evolve. Possessing an x-ray inspection system on a processing line could become an essential requirement for attaining status as a qualified supplier of food products and ingredients to industry and the trade in the foreseeable future.

7.2 A brief history of the discovery of x-rays and other governing principles of physics

Under the cover of darkness in November 1895, Wilhelm Conrad Roentgen was working late in his laboratory. His goal that night was to better observe the eerie glow produced by the electrical stimulation of a

Crooke's tube. In the latter half of the 19th century scientists like Roentgen were taking advantage of the newly developed vacuum pump to evacuate such glass-blown tubes and fill them with other various gasses. Electrodes were imbedded in these tubes, and the scientists would generate an electrical potential across the gap to generate various lighting effects referred to as "cathode rays" (11).

On the evening of November 8, 1895, Roentgen had settled into the darkness to better observe the glow of the tube. In order to gain a better perspective as to the mode in which the glow originated from the cathode, Roentgen covered the tube with opaque paper and stood back to make observations. As he sat in the dark room collecting his thoughts, he noticed flashes of light on a barium platinocyanide screen resting near the Crooke's tube setup. Roentgen was puzzled as to the origin of the mysterious light. The only source of light present, the tube, was covered from sight by the opaque paper. He disconnected the power supply to the apparatus and, to his surprise, the flashing ceased. After several iterations of energizing and de-energizing the tube to ensure the phenomenon was repeatable, Roentgen was convinced that some emission from the tube was the source of the flashes, and he called the emissions from the tube x-rays because of their mysterious nature of origination. In the weeks that followed Roentgen was able to determine that these x-rays could stimulate the flashing of the plate even when covered with dense materials like aluminum.

In 1901 Roentgen was awarded the first Nobel Prize in physics for his work on characterizing the phenomena (12). Amazingly, he never patented this unprecedented discovery and, to this day, the magnitude of the x-ray's impact on the world should be viewed as a gift from Wilhelm Roentgen (12).

7.3 How x-rays are generated, and why they can pass through materials of various density

In industrial and medical applications, x-ray generation is accomplished through similar means. Alternating current (AC) is supplied to an x-ray generator that controls the intensity of x-ray energy and length of the energy pulse. The generator is essential for supplying and controlling the power needs of the x-ray tube in a manner that provides x-rays possessing useful characteristics for radiography. The tube is essentially a glass or metal container possessing a high induced internal vacuum (1X10-5 mm Hg) (14). The high vacuum is necessary to ensure that there are few gas molecules present to interfere with the electron stream that will be generated when the tube is under an applied current. The remaining components of the tube generally present are:

1. A cathode filament made from a thin wire of tungsten that has provisions for two unique circuits.
2. A rotating anode consisting of a copper wire with a thin angled tungsten surface coating. This assembly is usually held in place via a graphite base.
3. A tube shield. The shield is usually made of lead with only a small port area that will allow the x-rays to escape in a prescribed direction. The shield functions to hold the tube in place and prevent undesired radiation leakage. There is usually oil circulated through the shield for cooling of the tube when it is under voltage.
4. Cable leads for connection of the anode and cathode to the high voltage circuit of the x-ray generator.

Figure 7.1 displays an illustration of a modern x-ray tube. The x-rays are created via a specific sequence of events. The necessary events are: (1) to generate numerous free electrons, and (2) to catapult these electrons at an extremely high velocity toward imminent impact with the awaiting tungsten target. The electrons are freed from the tungsten cathode via a process called thermionic emission (12). The emission occurs when the x-ray generator applies a low voltage, less than 12 volts, to the cathode via the filament circuit. This low voltage provides enough heat (energy) to essentially boil off some of the loosely held outer shell electrons of the tungsten compound.

The "freed" electrons at this point would leave the atom in any random direction, but in order for an x-ray to be generated it must be channeled to the anode with the awaiting angled tungsten target. This is achieved by applying a high voltage (up to 100 kv) between the anode and cathode that rapidly accelerates the electrons toward the awaiting tungsten target. The electrons will be sent toward the target in waves of various density and acceleration characteristics due to the nature of the alternating current used to generate the voltage.

When the electrons successfully collide with the tungsten target, there can be heat generated through ionization of the tungsten (accounts for 98 percent of the electron interactions with the tungsten target), or in a minority of cases other reactions that will produce x-rays occur. The main

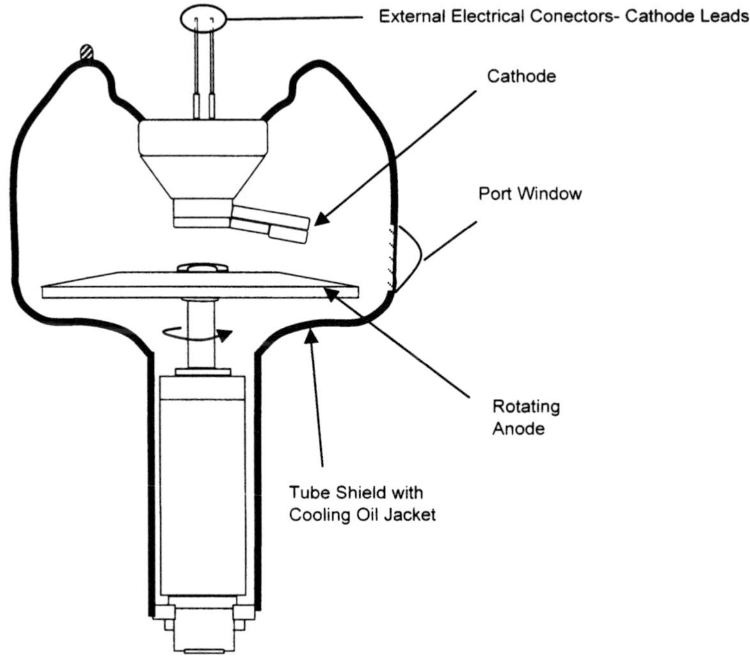

Figure 7.1. Sketch of the components of a typical x-ray tube.

mode of x-ray production during periods of interaction of electrons with the tungsten target is through a process called bremsstrahlung, a German word meaning "braking radiation." Essentially, the streaming electrons can travel at any path to interact with the tungsten target, but when they pass close to the positively charged nuclei of the tungsten, a so-called bremsstrahlung phenomenon can occur. Thus the path of an electron, now traveling at a very high velocity thanks to the high voltage, is influenced via the large attraction to the positively charged proton-rich nucleus. The electron will be attracted to the nucleus, and its path of travel will be influenced to a greater extent with closer proximities. The process is reminiscent of the real world situation of having to make an unexpected sharp turn at an intersection when traveling in a car at high speed. As the electron "makes the sharp turn" around the nucleus, it gives up a great deal of kinetic energy as it decelerates. This packet of energy, or photon, is emitted and is the source of the x-rays that puzzled Roentgen. Figure 7.2 provides an illustration of the process of x-ray creation through bremsstrahlung.

The x-ray spectrum produced through bremsstrahlung (tungsten in this case) possesses a normal distribution of radiation of all frequencies (en-

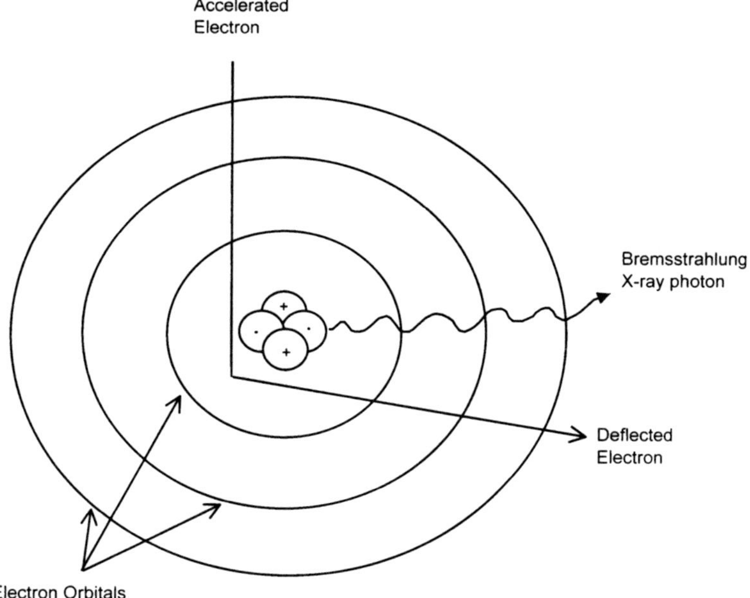

Figure 7.2. Production of an x-ray photon via bremsstrahlung process.

ergy) up to a maximum. The maximum x-ray energy produced is dependent upon the maximum voltage applied by the generator, not the amount of or distribution of the target material. For this reason the strength of x-ray fields are defined in KeV. Because the x-ray photons occur when the electron is taking a turn, so to speak, they will travel in various random directions in relation to the target material. For these reasons the tube shield has a window of various dimensions usually made of a low-density material, like aluminum. The window acts to weed out the weaker x-rays and guide the path of those x-rays of sufficient strength toward the targeted inspection area. If the x-ray beam is too wide, a mask or collimator made of lead can be used to further narrow the area of exposure.

7.3.1 Attenuation of x-rays

As previously noted, the path of x-rays can be blocked through use of dense materials, like lead, for shielding. The process by which radiation, such as x-rays, loses energy as it travels through matter is called attenuation. The attenuation of x-rays in materials placed in the inspection beam's path is what allows contaminants to be identified in the product. There are multiple ways in which x-rays can be attenuated, depending on how the photons interact with the matter present. The three basic forms of attenuation for x-rays in order of prevalence are:

1. Photoelectric absorption
2. Compton interaction
3. Pair production

Pair production is rare at the energy levels used for inspection of food products, so it will not be presented in this text. Photoelectric absorption and Compton interaction will be discussed briefly below to provide the reader with a thorough understanding of how the components in a product can attenuate x-ray photons.

7.3.1.1 Photoelectric absorption

X-ray photons can be subjected to photoelectric absorption processes as they pass through matter. Photoelectric absorption occurs when the x-ray photon collides with an electron that is bound to an atom. When this violent collision occurs, the energy released is sufficient enough to actually knock the electron completely free from its orbit. The probability of this occurring is tied to the probability of the photon striking a bound electron in an area. This fact makes it easier to understand why compounds made of atoms that have vast amounts of electrons (such as lead) can effectively block the flow of x-rays and conserve their energy (12).

7.3.1.2 Compton interaction

X-ray photons can collide with free electrons, or with loosely held electrons of compounds as well. When this occurs it is referred to as "producing Compton scattering effects." In this case there is a collision between the photon and the electron which results in the electron being energized with only some of the energy from the photon. The photon is still present, but is now traveling in a different direction than its original path. The photon is now deemed a Compton-scattered photon due to it now possessing less energy and traveling on a different path. This situation is very similar to the interactions that occur when one is playing a game of billiards. The cue ball will strike the target ball with momentum and send the target ball off in a new direction. The cue ball itself still continues to move, but does so now on a different path with reduced velocity.

Attenuation processes are wholly based on the probability of the x-ray photon being involved in some sort of collision. The photon travels at the speed of light toward, and potentially through, the object if its density is conducive to not fostering a molecular collision. The probability of attenuation can be mathematically modeled for different types of materials. As a general rule, solid materials containing elements of higher atomic mass will result in the greatest amount of x-ray attenuation. Liquids will vary in attenuation based on their elemental composition and amount of dissolved solids. Most common gasses (O_2, CO_2, and N_2), however, will provide the least attenuation effects. This is because the molecules in a gas are spread farther apart than they are when the material is a solid or liquid. Obviously the thicker the material envelope in the inspection zone, the greater the probability of attenuation. Imagine the tortuous path an x-ray photon will have to pass through when it meets a solid matrix of 10 cm-thick austenitic stainless steel. Now imagine the same photons traveling through a 0.4 mm mylar bag that contains pasta. The attenuation values for various materials at increasing levels of applied x-ray energies are presented graphically in Figure 7.3. The values for the graphic were obtained from Hubbell and Seltzer (reference 10 in "Works Cited"). Note the apparent differences in the attenuation of the materials at lower x-ray energy levels,

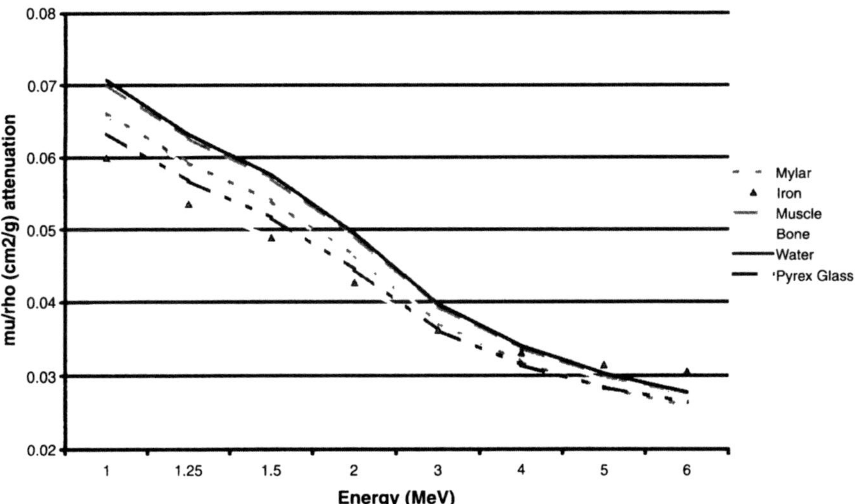

Figure 7.3. Graph of attenuation coefficients of different materials over a range of x-ray energies.

and how the attenuation differences among the materials decrease as the energy is increased.

7.4 How images are produced from x-rays

If all x-ray photons were attenuated when they met solid or liquid matter, they would be of little use in identifying foreign material in food products. The x-ray photons that pass straight through matter, or are mildly attenuated via Compton scattering, are the ones of interest in this case. These voids in the presence of x-ray photon distributions can provide a picture of where dense objects may be when the proper detector setup is used. The x-ray photons can be used to produce a picture of the product and its density variation, assuming that the x-ray energy applied to the inspection product is sufficient to allow a decent quantity of x-ray photons to pass through on a routine basis. The trick to this, however, is to ensure that too much x-ray energy is not applied to the target. Applying too large an energy to the target will make the differentiation between materials less apparent.

Certain materials, like the barium platinocyanide plates that were in Roentgen's laboratory, fluoresce (give off visible light) when x-ray photons impact their structures. This fluorescence phenomenon is used to visualize the remaining components of the x-ray photons that remain. The images generated from the fluorescent x-ray systems used in industry are positive images, meaning that as the materials are exposed to x-ray energy they give off more visible light and become brighter in color. Unlike medical x-rays, that are negative images where exposure to x-ray photons makes the film darker, these images have to be read differently by the person or computer.

The earliest systems developed for use in radiography of FM in foods utilized a screen coated with phosphor—a material that is highly fluorescent when x-ray photons are present. Essentially an operator would be positioned on the opposite side of the screen and view the product image through a thick section of leaded glass. The operator would look for dark spots, indicating FM contamination was present, and cull any abnormal containers as they reached the discharge of the machine. These systems used a high intensity continuous x-ray tube that saw duty in medical application. The typical energies of the era were greater than 120 KeV (12). Contrasting was achieved by increasing or decreasing the thickness of the phosphorus layer applied to the screen, and applying a lower voltage to the x-ray tube if necessary. The development of the image intensifier and other automated imaging techniques have removed the human from looking in the glass window to see the image, and have added a new dimension to analysis of x-ray images.

7.4.1 Image intensifier systems

The image intensifier was designed in the 1950s for use in medical x-ray systems. The image intensifier helped to aid in achieving better contrast of the acquired image and to reduce the blur that was present in x-ray films due to miscalculation of the photospot (14). The development of the image intensifier also created an environment where dynamic real time radiography was possible, and set the stage for automated inspection of food products for foreign material.

Basically the image intensifier is constructed within a vacuum-evacuated glass tube or metal housing. The housing has an input screen where the x-rays enter, and an output screen where visible light images are created. The material of the input screen is usually made of a thin aluminum layer ($<$0.5 mm) that is coated with a scintillating layer (fluorescence) and a photocathode or fiberoptic coupling. The scintillating layer is usually composed of CsI:Na, which has a higher x-ray absorption efficiency than phosphorous compounds used on the old screening technology. Each CsI:Na crystal can be as small as five micrometers in diameter, and they are deposited uniformly across the surface. The individual crystals, of small size, help to maintain excellent spatial resolution of the photon energy being received. The efficiency of converting the received x-ray photons into a light source is quite high in the CsI:Na matrix. The matrix will generate 3,000 light photons (at a wavelength of 420 nm) for each 60 Ke V X-ray photon absorbed (14).

The visible light generated at the scintillating layer is converted into electrons by the photocathode at about a 10 to 20 percent efficiency. These electrons are then accelerated toward the photoanode and focused via electrostatic focusing lenses. The resultant image formed on the output screen is an upside-down, left-right inversion of the original image that was captured at the scintillating layer (similar to the focusing present in the human eye). An illustration of a typical image intensifier is presented in Figure 7.4.

As previously noted, fiberoptic coupling of the emitted light from the scintillating layer directly to the output screen is being utilized in the

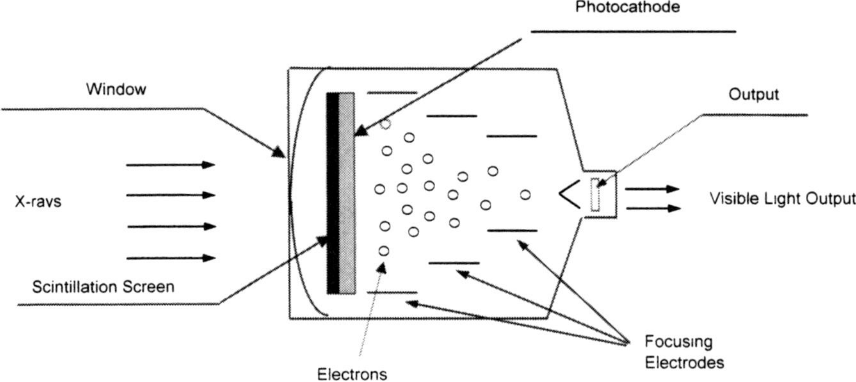

Figure 7.4. Illustration of the components of and function of a real time image intensifier.

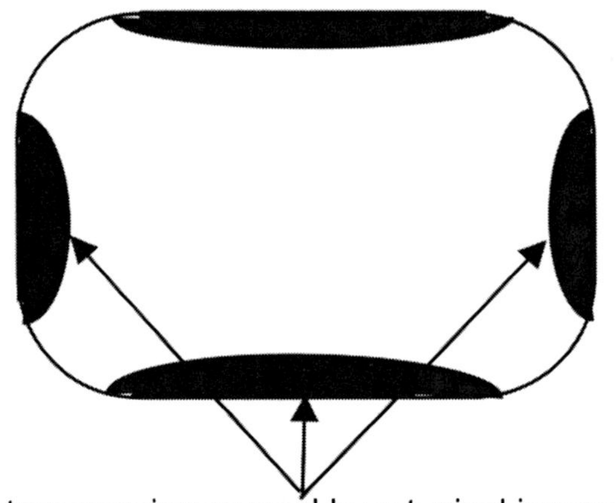

Picture warping caused by a typical image intensifier

Figure 7.5. Image artifacts encountered when using an image intensifier due to the focusing of the light.

latest renditions of the image intensifier. This has helped to reduce the image artifacts that have been common to all image intensifier systems. The image artifacts produced by the intensifier will suffer from a characteristic warping at the edges of the image (Figure 7.5). This is due to the fact that the light at the edges of the acquired image has been "funneled" and bent during the process of intensification.

Once intensified on the output screen, the image and artifacts can be read via any technology that can differentiate between the varying intensities of the light present. In the earliest days these systems utilized a closed-circuit television camera (CCT) to display the image on a larger monitor remotely located from the inspection area. Now, use of a CCD is common and allows for high resolution of the output from the x-ray image displayed at the image intensifier.

7.4.2 Photodiode arrays

Another way of converting the x-ray photons into visible light photons capable of producing an electronic image is via the use of photodiode arrays. Essentially, this arrangement utilizes a photodiode in place of the photocathode-photoanode coupling of the image intensifier. The photodiodes are coupled with either a fluorescent screen or possess their own scintillating layer in their lens to generate a visible light source from the x-ray photons absorbed. Typically, photodiodes used in these units are 0.7 mm × 0.7 mm in diameter or smaller, and can quantify received x-ray photons into electrical pulses quite well within that area. As in any other machine vision system, the electrical output of the photodiode is analog in format and must be converted to a digital scale so that a gray scale representation on a monitor, or digital image, may be recorded by a computer.

7.5 X-ray applications in the food industry and their components

The advancements in x-ray technology used in the food industry have come a long way since the early systems employed by Borden Foods and others. Today there are numerous configurations available from an ever-increasing base of suppliers for the processor to choose from. The various types of equipment that have been developed for use in the food industry, and their potential applications, will be discussed in various subsections below. A thorough review of each application is necessary for understanding the advantages and disadvantages of the type of inspection unit relative to each application.

There are numerous suppliers of x-ray inspection equipment who would be glad to sell an x-ray system to a processor, but one must ensure that it meets all basic needs before issuing a purchase order. Realize that there is a correct configuration for each application that exists, and selection of the type of system should be based on identified hazards ascertained via a thorough HACCP/QACCP review. The review should include not only known FM issues for your products, but those known to be an issue for your industry segment as well.

7.5.1 Static or rotating-image continuous phase bench systems

The static system offers utility in situations where an occasional product would need to be analyzed for defects or for assurance of proper construction off line. This type of inspection system is the most inefficient in terms of throughput, but can offer the greatest quality of image at differing perspective for analysis of the product (i.e., image resolution and magnification). Many systems of this type are used for NDT of packages to ensure that their construction meets established specifications. A bench system of this variety is especially helpful in the canning industry for identifying can assembly defects that could jeopardize hermetic seal integrity (false seams, cut-through of sealing compounds, etc.). When used in this manner, though not a deterrent to foreign material contamination, the processor has the opportunity to eliminate the tedious process of conducting extensive numbers of destructive can-teardown examinations.

A use of this system in foreign material identification/prevention can be found in the examination of product that is implicated in a tampering incident. By being able to thoroughly examine the container in a nondestructive manner, the processor, or regulator, may be able to determine if the contaminant was introduced at the facility or by post-process handling somewhere else in the distribution network. The advantage of performing the evaluation by using x-ray techniques in lieu of destructive testing is that the latter could damage crucial evidence for the prosecution of culpable parties, or further cloud the mode of tampering.

The components and construction of an off-line bench x-ray system vary with each application, but they can generally be categorized in differing categories depending on the presence or absence of the following features:

1. Fixed or variable source energy
2. Sample presentation—fixed or moving
3. Use of an image intensifier and camera, or a screen box
4. Data capture of image and computer enhancement versus as displayed

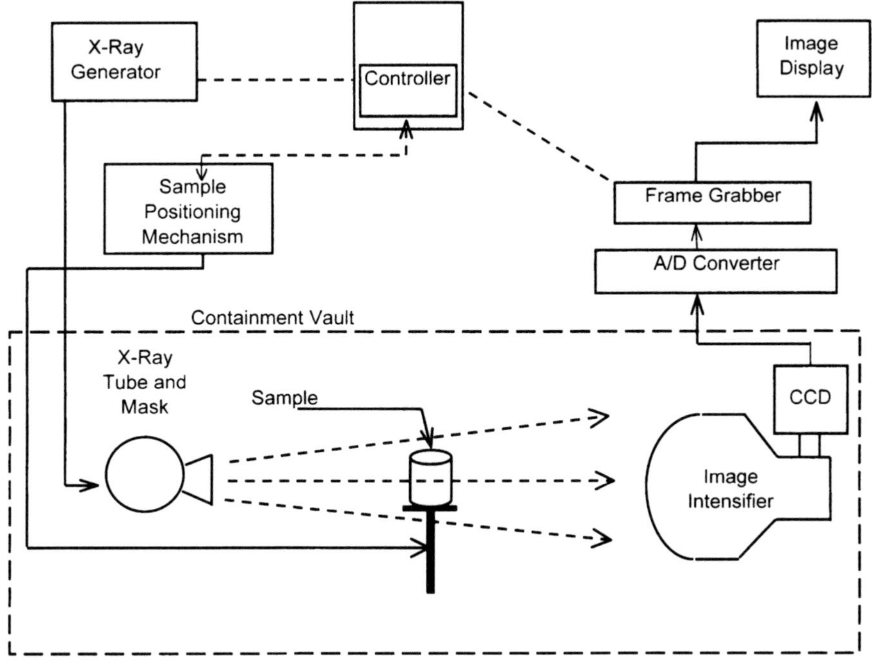

Figure 7.6. Typical components of a rotating continuous phase bench system.

The assembly of the components present in a typical off-line system is represented in Figure 7.6. The benefits of enhanced imaging, resolution, and finite control aspects of this system are outstanding, but the overall value of the system in preventing contamination of food products is quite limited in scope for the investment required. Hence, these systems should only be considered for the laboratory or research and development environment.

7.5.2 Offline, or slow inline, continuous phase systems

Sometimes offline systems to evaluate products are required by processors. A case in point would be a rework situation for suspect FM contamination of products, or the conductance of a validation study on a new packaging technology. Typically in such situations, the equipment would either be rented by the processor, or company purchasing the products, in response to a known failure. In some cases the product would be shipped to a third party that owns an off-line system for inspection prior to shipment on a contracted basis. Many times the use of these systems comes into play when an unforeseen situation occurs and the processing facility did not have inline detection methodologies present. Adverse situations dealing with supplied raw products or food packaging are typical drivers for processors to rent these inspection systems. Failed inspection equipment at the supplier location, or a container manufacturing issue that could increase the propensity for breakage, are usually issues that can cause a processing operation to lose face in the quality of the supplied product and seek inspection as a remedy. Typical examples of why finished products would be subjected to x-ray analysis include thin container sidewalls, presence of excess stone defects in glass containers, presence of loose solder in metal cans, extruder buildup, flecking off into a molded plastic container, etc.

The design of x-ray systems used for this type of inspection are varied, but they are usually based on an image intensifier design. Generally the product being inspected is conveyed into the shielded x-ray inspection zone via a conveyor belt. Depending on the application the x-ray source will be set up in either a vertical or horizontal configuration to the product, and a shutter-type collimator will be utilized to reduce scatter and isolate the field of view. The image from the image intensifier would typically be captured by a camera system. The output from the camera will either feed the signal to a computer for further analysis, or a monitor for discriminating review by an operator. The systems available are normally manned by an operator, who evaluates the image and makes the determination to accept or reject the product and actuate a rejection device if necessary.

7.5.3 Pulsed and continuous online image intensifying x-ray systems

Advancements in x-ray technology have been coupled with advancements in image acquisition and ever-increasing processor capabilities to deliver real time continuous inspection of products as they are manufactured. The core components of an online setup consist of the typical components found in any machine vision system coupled with the components necessary for generation and conversion of x-ray photons to visible light. The components necessary are:

1. X-ray generator
2. X-ray tube
3. Collimator
4. Image intensifier
5. CCD camera (area), or photodiode area detector array

The nuances of the continuous setup are found in the mechanisms used to convey the product, trigger the x-ray image capture, and analyze the images collected. A simplified illustration of a continuous image-intensifying x-ray inspection system is presented in Figure 7.7.

The pulsed x-ray has the advantage of being able to modulate the generation of x-rays in a manner that is coupled to the product inspection activity. In this scenario the x-ray generator is tied to a trigger, usually a fiberoptic photoeye, that will modulate the activation of the x-ray pulse. The pulse will correlate to a period of time when the product is in the inspection zone. This operation strategy means a much longer lifespan of the system components, especially the x-ray tube, and image intensifier versus

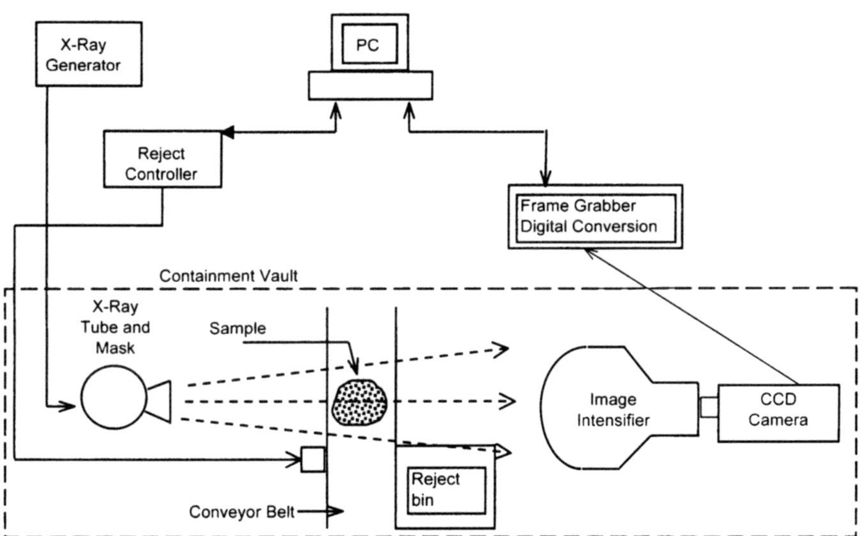

Figure 7.7. Schematic of the typical components of a real time continuous image intensifying x-ray inspection system.

a continuous operation. There is a bit of added control logic and components for this setup, but the minor additional upfront costs usually pale in comparison to the increased service life of the x-ray system components.

The hypothetical example of continuous x-ray inspection of a premium juice product packaged in glass bottles is presented below. This example should provide a sufficient understanding of how the technology can be used to identify FM within a product.

7.5.3.1 Case study—Get the edge! X-ray inspection at Jagged Juice Company

Jagged Edge Herbal Tonics manufactures hot-filled pasteurized juice products in glass bottles with metal screw-on closures. The typical process for Jagged juice is to run empty glass through an inverter washer, revert it, and run it to a large multihead filler. The product is filled to the required volume while hot (at 80°C), then conveyed to a rotary capper where it is capped. The glass received at the facility will sometimes contain foreign materials, and many times also possess defects that can promote breakage. Thankfully, the marginal quality glass was sourced at an affordable price that allows Jagged juice to enjoy a comfortable operating margin.

During the filling and capping process there is a tendency for glass breakage to occur and potentially contaminate other bottles within the area. Knowing that not all breakage will be evident to the personnel running the high-speed bottle-filling line (running at >700 bottles per minute), the company has installed an x-ray inspection unit. The unit was installed at the point of the process immediately prior to where the capped jars enter the cooling tunnel. This point in the process proved advantageous because, two meters farther down the line, the bottles are deposited on a wide mass conveyor and lose their single file arrangement. The bottles of juice then proceed to the x-ray unit via a single file conveyor where they are transferred to the x-ray machine. The machine uses side gripper belts to carry the juice bottles to the x-ray inspection zone. The speed of the side gripper belts is controlled via a variable speed drive coupled to the speed of the single file infeed conveyor.

When product is present on the conveyor, a photoeye will sense the incoming bottle and trigger the frame grabber to prepare for duty. As the bottle travels under the x-ray head, it travels over the input screen of the image intensifier. The frame grabber sequesters an image from the monochrome CCD camera monitoring the image intensifier output screen. The signals from the CCD are converted from an analog signal to a digital signal, and then assembled by the computer processor into a 256 gray region map. The map is analyzed in 256 gray mode using a thresholding algorithm.

Any containers that display a sufficient number of dark imaging pixels, greater than the threshold value, are rejected from the conveyor onto a metal tray adjacent to the exit conveyor. A photoeye is placed at the tray in a manner that the rejected bottle is accounted for, and the analog signal is sent back to the x-ray control unit for positive documentation of the reject. Operators routinely monitor the kickouts of the juice product by pouring the bottle contents onto a 30-mesh screen. The screen is washed by the operators, and any finds are collected and sent to the Quality Assurance Department for analysis.

When changing over to different product types (packaged in the same bottle and closure combination), the threshold algorithm that was determined as optimal for that specific product is loaded into the system. By having different programs for pulpy or clear juices, the number of false rejects generated is minimized while sensitivity for small glass fragments, potentially present within the products, are optimized. A well-marked seeded sample of the juice product is sent through the detector approximately every half hour by the operators to ensure the system is functioning as designed. The operators document all system checks and reject containers identified throughout the shift on a HACCP record that is reviewed by the Quality Assurance Department.

7.5.4 Line scan CCDs and linear photodiode array inspection systems without image intensifiers

Scintillated arrays of photodiodes placed in various arrangements can allow for adequate imaging of products at high speeds. Likewise, use of CCD line scan cameras placed behind a scintillator can provide for high-speed continuous inspection of conveyed food products. These types of systems have recently been used as successful replacements for pipeline metal detectors due to their ability to sense contaminants in conductive product matrices. There is no image intensifier used in this arrangement, and that provides for a direct pixilation of the product image with limited image artifacts being introduced. Spatial resolution is slightly diminished versus an image intensifier setup, but the simplicity of the system is a definite advantage.

The components utilized in these systems (minus the shielding materials, x-ray generation, and scintillation components) are very similar to those in a typical monochromatic machine vision system used for sorting (discussed in Chapter 6). The same imaging techniques can be used, and the systems currently available using this technology have proven quite capable. There has been a large increase in the prevalence of these systems because of their reduced initial cost compared to continuous image intensifier CCD systems.

The CCD line scan camera-based systems have excellent resolution and will provide outstanding imaging of products that afford sufficient contrast. The systems are relatively simple to install in an existing operation as well. The equipment footprint is quite small compared to the cabinet-type image intensifying CCD systems, and a normal conveyor setup can be used. A typical installation is displayed in Figure 7.8. As long as the product on the conveyor is not subject to tipping over, or to vertical oscillations when going over the detection strip, the images produced are quite good and dense contaminants can be readily culled. The use of imaging techniques has allowed these systems to provide excellent contaminant recognition in many products while ignoring common compo-

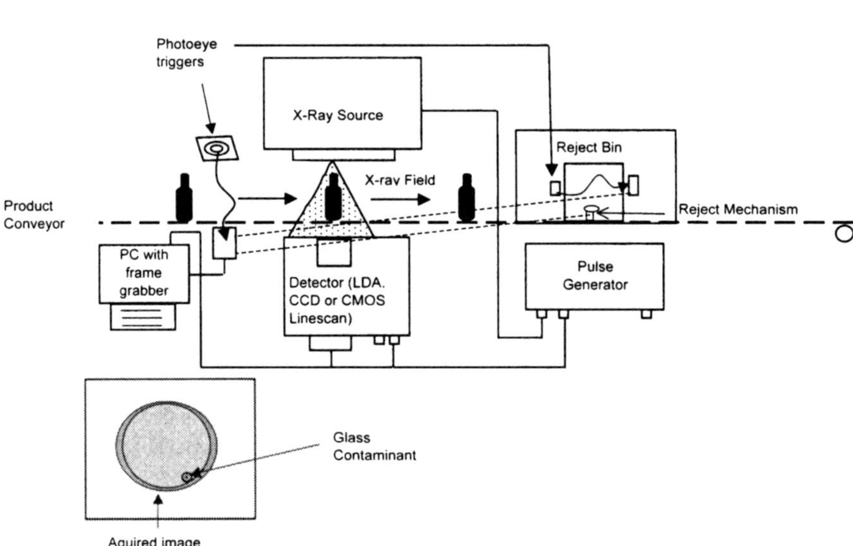

Figure 7.8. Schematic of the typical components of a real time pulsed line scan x-ray inspection system.

nents like soaker pads or other packaging features (pull tabs, foil films, and premiums).

7.5.5 Multiangle inspection systems

It is possible to use x-ray systems with multiple detectors and source paths to image a product in three dimensions. Collimation of the x-ray sources to specific detectors is quite difficult in practice; however, due to Compton scattering effects, the use of detectors in series would most likely be the best approach. The images generated will not be as clean as those of a typical visible light machine vision system, but the morphological information obtainable on FM contaminants and other product quality attributes is outstanding. Unfortunately the cost of multiple inspection components and the programming that is necessary to positively track and integrate the two images make such a system prohibitively expensive for most implementations.

7.5.6 Dual-energy imaging systems

Dual-energy x-ray imaging systems, though commonplace in medical radiology and security checkpoints, were not used in the food industry until recently. Graves and others have developed a novel x-ray detection system with a unique array of photodiodes that can simultaneously image the lower energy fractions of the incident x-ray photons as well as the higher energy fractions present. A combination of custom filter and scintillator materials block high-energy radiation above 40 KeV from being sensed by the first row of sensors within the array. The second row of sensors essentially filters out x-ray photons of greater than 20 KeV (16). The lower energy fractions allow contrast of components of various densities contained in thin low-density materials, while the higher energy signals can penetrate through thicker, denser materials to identify various structures and contaminants.

Novel software processing techniques and registration of the photodiode array via a unique CCD register sequence allow for an image to be produced that displays detail of both the high and low attenuation components in the same frame. A proprietary image subtraction technique isolates the area where the absorption characteristics of the product differ most widely. A complex neural network is used to discern whether these areas represent the contaminants that the system was initially trained to identify and subsequently reject. Current inspection systems are checking 20,000 chicken pieces per hour, and the products are guaranteed to be 99 percent bone-free (16). This technology has been marketed toward identifying thin low-density bones in fish, poultry, and other meat products, but could have far-reaching applications for detecting contaminants that are poorly contrasted using traditional x-ray technologies.

7.6 Automation of x-ray analysis

Automation of x-ray image analysis is rapidly improving, thanks to the ever-increasing processing power of computers and crossover programming techniques from machine vision systems. Image analysis capabilities have been extended tremendously due to fast input/output (IO) schemes being combined with high-speed processing and algorithms available for analysis of the digital feed.

There are currently two schemes utilized to identify a "bad" product in x-ray analysis. The techniques used in x-ray component image analysis are basic thresholding and morphological analysis.

Basic thresholding is the simplest and quickest form of x-ray image analysis; however, its success in identifying small dense contaminants or large mid-density contaminants within products that are inherently variable in density is marginal. In a typical thresholding setup, the variation of gray scale in a product is characterized in the same manner as a monochromatic machine vision system would be. Reference samples of non-contaminated products and contaminated products are presented to the inspection field, and the gray scale images are read. The variation within and among the gray scale components of the image are statistically analyzed, to model the variation, and a threshold point is chosen for operation.

Some threshold systems use adaptive thresholding, which allows the threshold value to slowly "float" based upon the product characteristics encountered over a number of samples. This is akin to the digital signal compensation mode of a metal detector attempting to zero out product effects. The function does help to reduce false rejects (which can be an issue when using these systems). The adaptive philosophy can also be useful if the product has a tendency to entrap air or to fluctuate slowly in density over time. If an adaptive thresholding system is to be purchased, the vendor must be asked for specific details on the adaptive algorithm utilized. A poorly written adaptive algorithm could potentially allow the product signal to be compensated for to the point where no false rejects are produced, but the sensitivity of the detector has fallen below minimum standards. For this reason the threshold algorithm should have an upper limit on how far it is able to "self adjust."

Morphological analysis is useful when inspecting complex products of varying densities (like soups and some snack foods). The ability to recognize the potential contaminant by its size, shape, gray scale distribution, or angular characteristics adds a new level of intelligence to the threshold scenario discussed previously. The most advanced systems use multiple-layer neural networks to better discriminate contaminants from products. These systems also have the advantage that they are more readily "trainable" to simultaneously perform and report the result of product quality inspections (8). The systems that incorporate morphology and shape recognition usually will represent a higher cost when compared to the thresholding systems available. Realize that the physical components of the systems, in principle, are the same, but the processor will be paying for programming and services that are not normally required in a threshold scenario.

7.7 Important points to consider when choosing a system

When purchasing modern x-ray inspection equipment, there are several factors that should be addressed by the processor prior to issuing a purchase order. A discussion of some of the more common points often botched by processors is presented below.

7.7.1 Identify where the inspection will be utilized, and what contaminants will be targeted

The processor must first identify at what stage of the process the x-ray inspection is going to be used, and what contaminant is to be isolated. Many processors are now using x-ray systems in place of in-process metal detectors. This is because of the x-ray system's enhanced ability to sense the highly austenitic grades of stainless steel (grades 304 and 316) while simultaneously identifying glass and some types of hard plastic contaminants. Applying a thorough hazard analysis methodology and QACCP review will guide this process better than a "yup, it would probably fit good here" method. It is important to ascertain if the area will be appropriate for the detector. Space considerations and easy access to the rejected containers are all questions that need to be addressed early in the selection process.

As stated previously, full-case x-ray inspection is available for use as well. Remember, however, that corrugated packaging materials can contain metal pieces capable of producing false rejects. Also note that, if the product is stacked upon itself (typically done in a case design), a higher amount of energy will be necessary to compensate for the additional attenuation presented by the product burden. This can result in the contrast normally presented for less dense and smaller contaminants to be lost in the image.

7.7.2 Identify the specific products and packaging configurations to be subjected to inspection

After deciding where and how the x-ray detection will be used, the next step is to identify the scope of the inspection. The number of different products and/or packaging configurations to be inspected should be gathered in this phase. The products should be categorized according to these factors as well as to their respective density and relevant particle distributions (if applicable). This information will be helpful when discussing applications with vendors, and can help to ensure that an overpriced inspection system is not purchased. Remember that extensive programming and preset memory features contribute to the overall system cost. If such features are not necessary for the specific application, there is no need to invest in them.

Samples of the products chosen for inspection should also be gathered before contacting vendors. These samples should represent the overall variation that the inspection equipment will experience at that point in the process. The vendor will most likely want these samples sent to them very early in the feasibility analysis, along with any drawings of the containers, closures, or case designs (if applicable) that detail the relevant dimensions and materials used. These items will be used by the manufacturer to develop components such as masks, software maps, and reject-tracking mechanisms.

7.7.3 Vendor service, credibility, spare parts, and preventative maintenance

When comparing applications offered by vendors, the ongoing service agreements and the quality of service offered can be a critical selection factor. A number of questions should be asked of the vendor at this point. Questions that can help to understand the validity of a vendor's service are:

1. Is 24-hour technical support available via telephone?
2. Can the system be set up for remote access via modem or internet to repair program glitches?
3. Will the vendor periodically send a representative to the site upon request for inspection of the system, and aid with any necessary regulatory reporting requirements?
4. Is the vendor in a financially sound position to maintain its current staffing levels and support staff?
5. Has the vendor recently been acquired by a new parent corporation, or recently been spun off from a company?
6. How long has the vendor been in business, and who else uses the system?

Remember that most x-ray systems available today are assembled with parts sourced from various suppliers. Just because Company X's name is on the detector does not mean that they built all of the components, or even assembled the system. An x-ray detection system is just like any other product, in that supplier relationships, manufacturing systems, and logistics are involved. The ability to quickly obtain original equipment manufacturer's (OEM) replacement parts and have them installed by a knowledgeable person is an outstanding perk when choosing a vendor.

This availability of parts can be especially problematic when dealing with unique x-ray inspection systems or low-volume producers. When solid state electronics are subjected to sustained doses of x-ray radiation, they will fail more rapidly than in other normal applications. When dealing with cameras and photodiodes, their failure patterns will become evident as the base image begins to get darker and darker with use. A significant increase in this dark current should flag a replacement being ordered, because the sensitivity of the system is being reduced. Image intensifiers and other components present in the field, however, should be carefully scrutinized for life cycle data from the manufacturer. The source tube is also an extremely important system component, and in time will fail without warning.

Experience dictates that these parts will fail at the worst possible time, so understand the vendor's spare parts stocking practices and shipping policies before purchasing a system. Also require that the vendor provide preventative maintenance procedures and recommended frequencies for performing them. This should ensure that the equipment is optimally maintained by the processor once it is purchased and installed in place. One part of the system that is often overlooked in preventative maintenance programs provided by the vendor is the host computer. Remember that the hard drive of the host computer provides intelligence to the system, and it is a critical component. Ensure that a system does exist for regular back up of the hard drive to a spare drive stored elsewhere within the facility. If the hard drive in the computer for some reason becomes corrupted, or fails, having an inexpensive back-up readily available to get the system up and running again is great insurance. Obviously, if the detector was purchased to address contamination issues deemed likely to occur, then the notion of manufacturing products without the unit should not be a very appealing alternative for the processor.

Works Cited and Select Resources

1. Chang R. 1994. *Chemistry. 5th ed.* Arlington, Tex.: McGraw-Hill, Inc. 1,088 p.

2. Khan A. 2003. *Field Reports: No Need to Roll the Dice When Inspecting Spice.* Food Engineering (2):online. Available from: <www.foodengineeringmag.com>. Accessed Oct 28, 2003.

3. European Commission 1997. *Finding a Needle in a Jar.* European Commission—RTD Info (15):1.

4. Safeline AVS. 2003. *Food Safety and the Role of the X-Ray Inspection.* Available from: www.packworld.com. Accessed Oct 25, 2003.

5. Gregerson J. 2001. *Tech Update: 20/20 Vision.* Food Engineering (6):online. Available from: <www.foodengineeringmag.com>. Accessed Oct 28, 2003.

6. Hardin W. 2003. *Visible to X-ray:Food Packaging Uses Entire Vision Spectrum to Meet Its Needs.* Available from Machine Vision Online: <www.machinevisonvision.org>. Accessed Oct 30, 2003.

7. Hartman LR. 2001. *X-Ray Inspection Boosts Confidence Through the Pipeline.* Packaging Digest (10):54.

8. Higgins KT. 2002. *Technology Update: Double-duty Inspection Tools.* Food Engineering (10):online. Available from: <www.foodengineeringmag.com>. Accessed Oct 28, 2003.

9. Higgins KT. 2001. *Refinements Coming to Packaging Line Detectors.* Food Engineering (9):online. Available from: <www.foodengineeringmag.com>. Accessed Oct 28, 2003.

10. Hubbell JH, Seltzer SM. 1996. *NISTIR 5632—Tables of X-ray Mass Attenuation Coefficients and Mass Energy-Absorption Coefficients.* Washington, D.C: Natl. Inst. of Standards and Technology.

11. Kotz JC, Purcell KF. 1991. *Chemistry and Chemical Reactivity. 2nd ed.* Fort Worth, Tex.: Saunders College Publishing. 1,066 p.

12. Martin JE, Lee, C. 2002. *Principles of Radiological Health and Safety.* Hoboken, N.J.: John Wiley & Sons, Inc. 790 p.

13. Pehanich M. 2004. *High-tech Inspection Enters the Mainstream.* Food Engineering (4):online. Available from: <www.foodengineeringmag.com>. Accessed Jun 1, 2004.

14. Pettersson H. 1998. *Physics, Techniques and Procedures.* In: The Encyclopedia of Medical Imaging. Vol. 1. Lund, Sweden: NICER Institute. ISIS Medical Media. P 464.

15. Vachtsevanos G, Daley W, Heck B, Yezzi A, Ding Y. 6–8 Nov 2000. *Fusion of Visible and X-Ray Sensing Modalities for the Enhancement of Bone Detection in Poultry Products.* Proceedings of SPIE Photonics East, Biological Quality and Precision Agriculture II, Nov 6–8, 2000. Boston Mass.

16. Wilson A. 2002. *X-ray Imaging Checks Food Purity.* Vision Systems Design (5):online. Available from: <www.vsd.pennnet.com>. Accessed June 1, 2003.

CHAPTER 8

Proper Initial Validation, Ongoing Verification, and Change Control for Separation and Detection Equipment

8.1 Introduction

The process of controlling FM contamination does not end by merely going through the motions of purchasing separation and/or detection equipment. Validation and verification of their effectiveness and effect on subsequent processing systems are essential. Performing an initial documented validation study and conducting ongoing documented verification activities are the best practice for a processor to ensure that its FM defenses will remain robust. The validation and verification concepts that are presented in this chapter are not the rigid and rigorous types that the FDA requires of medical device and drug manufacturers in 21 CFR Parts 210, 211, and 820. Moreover, the concepts presented are such that they serve the purpose of error-proofing equipment purchases and installations while formalizing what "acceptable" performance is for ongoing operations. Equipment that has been proven with extended use, is a minor contributor to the overall FM program, and/or is impossible to develop criteria for, can be exempted from the process. (This should be a rare exception.)

All equipment, however, will fail with extended use over time. Understanding this concept allows the opportunity to plan ahead for the inevitable before failure occurs. Employing preventative maintenance (PM) of processing equipment, separation equipment, and detection equipment is a must, but emergency maintenance procedures can be well defined in advance of acute failure. This chapter is dedicated to providing the reader with a systematic approach to ensuring that the equipment infrastructure of the validated process remains an effective barrier to FM.

8.2 Why go through the trouble of expending resources on validation and verification strategies?

The installation of detection or separation equipment such as metal detectors, x-ray inspection systems, screening operations, and magnetic separators usually represents an expenditure of significant cost by the processor. The expenditure will be made in either a proactive or reactive manner for a variety of reasons to negate potential FM contamination. Most additions of detection or separation technology can be summarized into one of the categories listed in Table 8.1. As such, detection equipment should be given all measures of due diligence to ensure that its operation meets the original expectation.

The concept of investor satisfaction is quite applicable to the purchase of expensive detection technologies. If the equipment is purchased at a high cost for an application that a hazard analysis indicates as a borderline need, there will be high expectations for success. When systems of this nature fail or prove to be a detriment to processing efficiency that wasn't identified in the cost benefit analysis, there will surely be a higher hurdle of scrutiny by the body that is providing the capital when the next equipment purchase must be justified. Following regimented science-based validation and verification strategies will help ensure that the technology performs as expected in periods of operation. It is important to note, however, that every installation is unique—the concepts will be the same, but the testing will usually not be of a "cookie cutter" fashion.

8.3 Who is involved in the validation process—is it just a QA thing?

Replicating the concepts in this chapter will require input from internal and potentially external experts. These experts usually are not PhDs; in-

Table 8.1 Drivers for the purchase and installation of equipment to detect or separate foreign materials from food products, and in what mode of operation the management is acting when doing so.

Reasons for Installation of a Foreign Material Detection or Separation Device	*Mode of Operation Taken by Management*
1. To improve upon the current state of food safety or quality from an unsatisfactory status quo	Reactionary
2. As a response to an existing customer's requirement	Reactionary
3. As a means of displaying enhanced due diligence when a processor is implicated as being in a gray area of compliance by a regulatory agency	Reactionary
4. When a scientific hazard analysis is performed (HACCP or QACCP) on an existing process and indicates that a hazard (quality or food safety) may be reasonably likely to occur	Reactionary
5. When trying to garner new business, and the technology represents a sustainable competitive advantage over other competitors	Proactive
6. Exactly duplicating a process for a start-up operation that has been validated to be adequate at controlling the FM concerns in an existing facility	Proactive
7. Responding to the ever-changing body of scientific knowledge, and/or incidences that are effecting other industry segments—unforeseen events	Proactive or Reactionary

stead they are employees who deal with the processes and ingredients on a daily basis. These knowledgeable individuals should be identified as key resources because they possess the data necessary for the design of outstanding validation protocols. This simple fact applies to HACCP and QACCP plan development as well. Of the several hundred validation and HACCP meetings that I have either facilitated or participated in, I have yet to note one where I failed to learn something new about the process or ingredients from these experts.

It is a foolish mistake to identify testing conditions and success criteria in a vacuum of "experts by credentials" employing various modes of food safety group-think. Certainly there is an absolute need for the quality and food safety professional to be present and oversee the selection and validation processes for new separation and detection equipment. A cross-functional team composed of operations, engineering, and maintenance personnel, however, will provide the best chance for initial and ongoing success of the installation.

This team should be led by a food safety or quality professional who possesses finite knowledge of the food safety and quality requirements for the product and ingredients in question. The rest of the team members should be a blend of management and supervision who are willing to share their experience and data relevant to the equipment selection and validation process. The most important characteristics that the other team members must display are a willingness to think critically and to listen generously to other team members who may not be as articulate or as comfortable in speaking in a group setting. The total number of team members should not be excessive; generally groups of fewer than nine people are able to be most effective. The team must be made aware up front that their goal is to construct a validation approach for the system that resembles Figure 8.1.

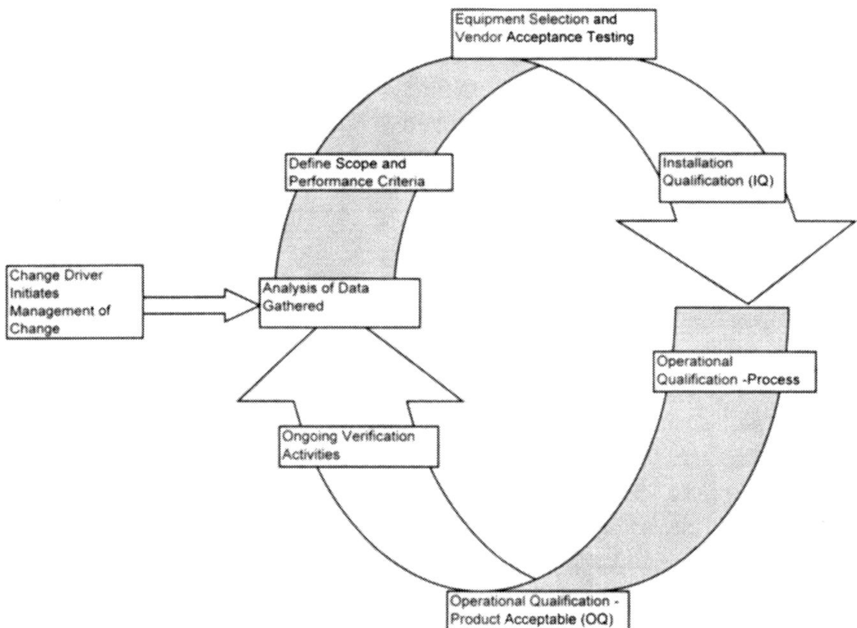

Figure 8.1. Representation of a living validation system for use on existing critical processes, retrofit of existing processes, and installation of new process technologies.

8.3.1 Validation—Step 1: defining the scope

Detection and separation technologies represent a unique variety of equipment that integrate electrical, mechanical, and pneumatic systems with a control scheme to identify and/or cull FM from edible food products or their subcomponents. The control schemes utilized can be intermittent and human-based in nature, or they can be real-time automated systems utilizing feedback control loops and complex neural networks. The previous chapters of the text can help to identify which technologies would be the most advantageous in specific situations, but the producer must determine the desired efficacy of the equipment to be utilized. By performing this task the processor will have developed an initial scope that outlines the minimal functional requirements that must be met.

It does not matter if the equipment is going to be installed in a new processing operation or a previously existing process—there must be documented criteria developed that outlines successful performance. Without written criteria how else would anyone involved with the project or operation know whether or not the equipment is working "as designed"? A typical scope of design will detail not only the desired contamination removal or detection performance of the equipment, but several other factors as well. Essentially, the scope should be a written description of how the equipment will ideally operate when "living" in the processing environment over a period of time. Factors usually contained in a typical scope include material of fabrication, supply date, contaminant removal at maximum throughput rates, false reject or scrap generation, documentation requirements, products to be processed using the equipment, machine acceptance criteria, payment schedules, service contracts, spare parts purchased up front, minimum definitions of cleanliness, suitability for operation in the environment, and utility requirements. Going through the process of writing a scope is a value-added process and, when finished, the validation team is a great way to get everyone in conceptual alignment.

8.3.2 Equipment selection, vendor selection, and acceptance testing

Once the team has reached conceptual alignment on what the detection or separation technology must look like to meet the expectation, it's time to go shopping. Use of sound theoretical and practical engineering and operations principles are a first step in the assessment of purchase options. It should be noted at this point that a source of funding needs to have been identified before engaging in lengthy discussions and testing protocols with potential vendors. The monies need not be fully approved at this point, but there should at least be a budget or a capital appropriation pool containing funds that could be diverted to purchase the equipment. Nothing is more disheartening for the vendor than to go through a lengthy process that ties up resources only to hear that there isn't money available this year, and there might not ever be any.

It is acceptable at this point to ask any potential vendor of a technology solution for references and the number of current applications that are in use. Written testimonies are a good "feeler," but if the technology cannot be witnessed in operation then the amount of uncertainty is a bit high for spending comfort. It is important to gain the perspectives of others by talking to individuals who are in operations or quality assurance functions at companies using the technology.

One way to ensure that the technology the vendor is offering could provide real value to your corporation is to ask the vendor to arrange a site visit at another company using the technology. Obviously it would be a conflict of interest for a vendor to take you into a competitor's facility, and a vendor who is willing to do this should be discarded from consideration immediately. Visiting a producer that is in another segment of the food or pharmaceutical industry, however, is entirely acceptable and to be encouraged. A personal visit by team members can provide a good venue for evaluation of the technology's potential application in the home facility. Discussions with groups using this technology should go deeper than the typical "works good and feels about right" conversation. Ask for data on throughput, false rejects, maintenance requirements, ease of operation, vendor service quality, and overall reliability. Ask if they had a formal validation strategy when the equipment was installed, and what the purchase contract specified. Be sure to note any incompatibilities identified in the areas of sanitation, operation, and vendor service, and attempt to hold conversations with the employees who use the equipment on a daily basis. As previously stated, they are the true experts and usually will give you unfiltered opinions about the equipment. The data collected will be very important when negotiating with the vendor and constructing the functional scope of the purchase agreement.

Once potential vendor(s) have been identified, the team should attempt to tie the scope that was developed into the contract for purchase. Moving forward with the purchase and installation of any detection or separation technology without a written scope is a poor decision. Most reputable vendors, in fact, would not be supportive of a purchase agreement that does not include performance and acceptance criteria. This process will involve concessions being made by both parties and documentation of disclaimers about several unknown factors, but the buyer must stand firm on documentation of critical performance requirements in the purchase agreement. There will be times when certain requests, such as zero waste or self-cleaning equipment, are unjustified due to the limits of technology and overall control of the operation. Just use common sense, tact, and your knowledge of practices at other locations when negotiating through such sticking points.

8.3.3 Acceptance testing performed at the vendor location

The final product of the contract negotiation should resemble a stratification of success and payments to ensure that value entitlement is achieved by both parties. The core milestones that must be achieved, whether they are specified in the contract or not, are to ensure that the technology is successful, easily measured, and repeatable. The first hurdle that usually presents a performance-based assessment is the acceptance testing done at the vendor. Simply put, if a vendor cannot demonstrate that a technology will work in the confines of their test facilities, you do not want to pursue it further. The acceptance testing performed at the vendor site typically involves sending a range of products, representing ideal to worst-case scenarios, that the process occasionally will present to the technology for testing. Additional testing of equipment design, sanitation, environmental compliance factors, and personal safety requirements required by the Occupational Safety and Health Administration(OSHA) can also be performed at this time. These tests should be defined to grueling standards because the test conditions will be known by and in the control of the vendor's best technicians. The criteria for the acceptance test must be documented with clear limits on acceptance, and when the outcomes will be defined as a failure. A representative from the company should be present during this period to observe the results and record the data. In the event that a company representative cannot be sent, thorough documentation of the actual test conditions, results of the study, and video recordings of the technology should be requested. If any test or challenge shows that the equipment does not perform within the agreed-upon parameters, an evaluation should be performed to identify the cause of the failure. Corrections should be made and additional test runs performed, as needed, to verify that the equipment performs as required. If additional testing is unwarranted, then qualification and selection of a new vendor should ensue. The observed variability of the equipment between and within test runs, whether they are successful or not, can be used as a basis for determining the total number of trials and samples necessary for future qualification studies of the technology.

The documentation of these results, and any other acceptance test results that follow, should be performed following the generally accepted practices used for recording any other type of process data in regulated industry. This means that these documents should be written in a manner that would allow them to be subpoenaed for court or reviewed by a regulator in the event of a product contamination incident. Minimally, the record should: (1) be maintained in non-erasable pen, (2) identify the location of the test and the equipment used, (3) identify the name of the person observing and recording the data, (4) reference the time and date of the test(s)' start and stop times, and (5) document the results in a manner that is traceable to 1 to 4 above. If a computer-based data collection system is used, the results should be printed immediately and affixed with the information from points 1 to 4 above for maintenance in the paper record. Electronic records are convenient, but not recommended as an ongoing archive of this type of testing.

8.3.4 Installation qualification (IQ)

If the vendor acceptance testing was successful, the next step is to start tying in the new equipment to the process and getting all of the items in order for a successful installation phase. The installation phase should include a thorough review of the "as built" equipment to ensure it meets the agreed-upon specifications. Shipping mishaps can happen, product numbers can be transposed, and in-transit damage can occur so an inspection of the equipment supplied by the vendor could prevent an embarrassing and costly retrofit of an existing process. During the installation qualification phase, the team can review and/or write pertinent SSOP's, maintenance procedures, approved spare parts lists, and define the calibration procedures and frequency for the new equipment. Identification of operators in need of training, and subsequent provision of documented training to those individuals is also a requirement of the IQ phase. All of the training, maintenance, and sanitation procedures developed in this phase will need to be placed in a controlled document format and added to an audit schedule to ensure they are incorporated within the current quality or HACCP system.

The maintenance and sanitation groups should be allowed to practice their procedures during IQ, and offer suggestions for improvement. Emergency maintenance and sanitation procedures should be developed by these groups in addition to their existing standard operating procedures. Such procedures should include how to maintain, recalibrate, and clean the equipment when an unforeseen event occurs such as a breakdown or power outage. The procedure should include the steps necessary to bring the equipment safely back online without having the procedure itself be the cause of malfunction and subsequent contamination. The procedure(s) developed should actually be used in the IQ phase, and observed as a check of its true effectiveness. Such results should be documented in the same manner described in the vendor acceptance testing section.

The final stage of the installation qualification is to perform HACCP and QACCP reviews based on the new process description. Members of the validation team must be present at this review, and should bring "as built" drawings and factual process descriptions. Once the HACCP, SSOP's, maintenance, environmental, and personal safety systems have been reviewed by the appropriate personnel, the official shakedown of the system can begin.

8.3.5 Operational qualification (OQ)

Operational qualification of the new process possesses two requirements that must be met:

1. The process description, operational settings, and line speeds must be defined as formal operating procedures prior to any testing being conducted.
2. The ingredients used in the testing must have written specifications, and be characterized using a documented test method before and during production.

This may sound simple, but in truth there is usually a great deal of work that needs to be done up front to comply with the first tenet presented. If the processor already has a well-documented quality and/or HACCP system that is audited on a regular basis, then achieving this prescribed level of scrutiny will not be very difficult. If the processor's quality system is relatively new or lacking in conformance, it will require a great deal of effort to get to the point where Part 1 is achievable. The most fundamentally important concepts of Part 1 are that each process step is defined and described in a manner that is documented, subject to change control, and easily understood by the employees who will have to operate the equipment and perform ongoing testing.

For the characterization of the material, the reader is referred to the acceptance sampling chapter of this text (Chapter 9) so that the process can be done in a manner that provides statistical confidence. Again, with any other phase of the validation process, the documents should be kept in an official manner and placed in the hard copy validation record. Keep in mind that this phase of the OQ process can be completed quickly if a systematic approach is used by the team. In fact, a lot of the pre-work for the OQ process can be conducted parallel to the activities of the IQ phase.

The stage of OQ where the first product is produced that could potentially be saleable is at the product purity acceptance phase. Any product produced on the newly installed equipment prior to completion of this OQ phase should not be shipped to the trade unless 100 percent inspected by an equivalent process. If no inspection is performed on the product and it is released as saleable prior to completion of OQ, the manufacturer is assuming an unquantifiable risk. During the OQ phase the process will be challenged to assess its adequacy. This phase should consist of testing everyday process conditions and worst-case scenarios of throughput surges and incoming product quality. The product that is processed must be evaluated to ensure that its purity and quality are acceptable. This evaluation can take the form of 100 percent inspection, or be composed of statistical sampling throughout the duration of each test condition (see acceptance sampling chapter). Either technique will require the use of a documented test method that the processor believes is valid. Each test condition in the plan should be replicated at least once to ensure repeatability. From a scientific standpoint, three replications are desirable at each test condition to ensure repeatability, and repetitions using different personnel may be necessary to ensure that operator effects are screened out (3).

Involving a large number of operations personnel in the OQ phase is recommended because intimate knowledge of the new process is gained as well as invaluable exposure to the scientific approach used to ensure that the equipment functions properly. Operators who understand *why* a detection technology is in place will have a higher level of ownership than those who just know that it is supposed to be "on" when the quality staff is in the area. Knowledgeable personnel will be the first to defend the necessity of the equipment for continued operation when challenges occur on- and off-shift, or in a remote area of the operation. Such enlightened individuals will also have a good feel for when incoming product quality or throughput rates may exceed the design validation. The presence of the equipment manufacturer at the OQ is a must. By having the vendor on-site, the process of understanding and correcting any failures will be shortened. The vendor will have a vested interest in seeing the equipment perform successfully, and will usually be quite helpful in qualifying operators and providing insight from other start-up activities they had assisted with. In the end, the OQ process must meet the established acceptance criteria and the results must be appropriately documented. The only exception is when there is scientific justification to reduce the standard to an acceptable level that will not put public health, or the brand, in a situation of undue risk.

8.4 Ongoing verification of equipment effectiveness

Needless to say it would be a horrible waste of effort to go through the trouble of validating a process only to have it be operated in a manner that was never intended. Unfortunately this happens more often than could be imagined in food processing establishments around the world. Any true system is defined as such because there are a series of checks and balances. The cycle of continuous improvement—*plan, do, check, and act*—must be adopted for ensuring that the control mechanisms do not drift.

Conducting an audit frequency verification of the maintenance and operation of equipment is often the missing link in most foreign material control programs. Some technologies, however, are traditionally over-verified in certain circumstances, and this is driven from a lack of understanding of the equipment. Collecting data on the mean time between failures can help to guide the processor in establishing verification frequencies that better utilize resources. Remember that verification goes beyond just looking at the magnet or the reject mechanism, and should involve observations of the human elements of the process as well. If there is a human element that is required to ensure that an FM control technology remains effective, it needs to be observed (audited) on a frequency schedule that is probably at parity with the equipment itself. Human errors can have far-reaching effects and can result in the total failure of equipment and systems that prevent FM contamination.

8.4.1 Producing and handling of seeded samples and test cards

I once witnessed a major customer put a standard 2.5 mm ferrous test wand on a belt during a routine audit, only to see it go undetected as it proceeded unabated toward the next process step. This obviously was a dark day for the manufacturing facility, and the customer dropped the company as a supplier a few days later, citing a lack of confidence in the quality system. That simple exercise cost the company $130,000 in annual profit, and it took the supplier two years of courting coupled with a large unit price decrease to get the customer back. The saddest part about the whole exercise is that there was no standard operating procedure for verifying the detector's function. The detector was referenced in the company's HACCP plan, but the verification activity was not an official procedure or a controlled document. The employees working on the line would place a metal key chain under a pouch of product and, if it was kicked off, they recorded the detector as "functioning" for their check. Obviously a key chain is not an appropriate standard reference, and it surely does not meet the stringent requirement of being able to display that the detector could identify a 7 mm fragment of metal. When the operators in the area were approached to explain why this method was adopted, they were quite honest and helpful. They had simply lost their test cards some time ago, and were fearful of retribution if they were to ask for another from the QC department. Obviously this company had deeper issues than a nonfunctioning piece of inspection equipment or lack of document control.

Hopefully the case study of this supplier stirred up some questions, and exhibited disbelief of the situation. The true question that must be answered is what is an appropriate method of testing a detector? Many times the answer is less than obvious, but the key chain test certainly is not the correct way. In order to challenge detection equipment appropriately, the processor must work with the equipment vendor to identify a source of certified seeds. This is usually quite easy to do for metal detectors and X-ray inspection units. Most companies have multiple laminated cards containing ferrous, nonferrous, and stainless steel spheres of various dimensions. Samples that are imbedded in straw-like tubes are also usually readily available for purchase. Glass spheres of various composition (flint glass or pyrex) are also available for x-ray units, and spheres of hard plastic can be obtained if the vendor is harassed enough.

8.4.1.1 Creating the seed product and presenting it to the detector

The manner in which the product is seeded is very important in providing assurance that it is truly representative of the expected type of contamination. A good example of this is when attempting to identify glass contamination incidents that occur in a liquid product packaged in glass. The product itself usually does not provide a viscous-enough matrix to suspend the tramp glass particles (unless it is a heavy oil or puree). The glass particles would settle to the bottom of the bottle in a short period of time and the glass will usually be found in the heel of the container upon inspection. This is due to a concave bottom. A good strategy for replicating this contamination strategy would be to use a spherical 2.0 mm bead housed in the end of a straw. The straw could be placed in the sample container so that the bead would stay in the heel of the bottle for presentation to the inspection equipment.

For metal detectors, a good strategy is to attempt to model a worst-case situation. Knowing that the weakest detection field for a metal detector is at the center of its aperture can help in modeling this doomsday scenario. The sample should be presented with the test sphere(s) oriented in a manner so they will pass through this part of the detector. Straws for beads or placeholders for the key cards can be made that will travel on the conveyor and appropriately orient the sample. The test should be done with the product surrounding, or embedding, the test sample so conductive product effects can be modeled as well. Random presentation of "free" samples to detection equipment is fine on occasion, but it should not be the official mode of operation for verification purposes.

The previous general discussion on seeded product samples needs to be augmented with some common sense safety precautions. When the systems are properly developed, use of seeded product should never become a potential hazard for the consumer. All test containers/product that is to be placed on a running process line should be identified in a manner that affords easy identification. Bright fluorescent-type colors or a color that would never be present in the product itself should be used to provide the necessary visual contrast. This is extremely important because, if the reject mechanism fails and the product is on a high-speed conveyor, it could easily be sent on to the next processing step. Seeded containers should never end up in a consumer's hands as a "prize." There are few moments worse than seeing a poorly marked seeded container head unabated toward a pasteurizer or mass accumulation conveyor filled with 60,000 other containers that look just like it. Greater assurance for noncontamination will be achieved by putting the appropriate steps in place for the operational verification procedure, and contrasting the seeded containers appropriately. One example of a truly closed loop system is one where the seeds are uniquely numbered and issued from a central location, such as the Quality Assurance Department, to the line operators on a shift-by-shift basis. At the end of the shift the operator should return the samples back to the checkout location where a count will be made. If the issued count is not the same as the received count, some tough questions lie ahead for the operator.

Seeded samples should not be allowed to become a source of biological contamination for the product or in itself represent a biological public health hazard. They too need to be cleaned, rinsed, and sanitized if their surfaces are meant to be in contact with the food product. This is especially true when dealing with process lines manufacturing ready-to-eat products susceptible to biological contaminants (lunch meats, diced vegetables, potato salad, etc.). An instance of special concern would be when a seeded sample representing a biological public health hazard may be found when the seed is put in a container with product and hermetically sealed. This is usually done for producing representative seeded samples of canned products with particulate components or viscous liquids to challenge x-ray systems. Products such as soups, processed meats, oils, fruits, or vegetables usually require this kind of seeding. These products are seen differently by the x-ray detector than "dummy samples" filled with water or air because of the attenuation coefficients of the normal products being higher than that of the water or air matrix. This special case usually requires the seed to be placed in the "real" product matrix to ensure that sufficient sensitivity is achieved. Depending on the water activity and pH of the product, numerous types of pathogens may proliferate including *Clostridium botulinum* (if initially present in the product, seed, or packaging). The seeded containers should either be rendered as an unsuitable media for growth via: (1) acidification, (2) some form of chemical preservation, or (3) application of a thermal process (determined by a recognized process authority) to render it commercially sterile.

8.4.2 Verification of detection equipment operation

For detection equipment there are multiple verification steps that need to be employed. Not only is there a rejection and detection component that must be addressed, but usually a program can be audited as well. Such systems are complex, and the final settings that were used in the operational qualification must be documented in a manner where they are readily available for use. When auditing electronic systems, there should always be emphasis on ensuring that the entire system is working as designed, not just a single component or reject mechanism. Simply passing a seeded container through the detector may provide evidence that it is functional, but doesn't give assurance that it is operating in an optimal state. This is why it is important to verify that the programming has not changed. Sensitivity parameters have a tendency to be adjusted when product effects are persistently resulting in kick-outs. Adjusting the sensitivity of the detection equipment is an appropriate strategy for managing through a short-term product-effect issue; however, the changes that are made need to be officially documented and reviewed by a designee in the Quality Assurance Department for ongoing operation.

All detectors that incorporate a reject mechanism must be challenged completely. Most failures in detection systems are not at the point of detection, but at the rejection unit itself (3). In order for an operation to continue, the verification activity must include that the correct product was rejected. Many times low-system air pressure, maladjusted photoeyes, worn rejector plungers, wandering conveyor drives, or various other failures of mechanical and electrical systems can result in a missed reject.

8.5 Structured management of change process

Embracing a management of change philosophy for the equipment components, equipment settings, operational challenge procedures, and seed sample preparation is appropriate due diligence for any processor. This is especially true for processors who manufacture multiple product types, or designate the detection equipment as a critical control point in their process. A management-of-change process works by requiring any person who wishes to change the settings on a piece of detection equipment to seek approval before doing so. The change request would be formal, and documented in a thorough manner. The change would be reviewed by a committee consisting of management responsible for quality assurance, maintenance, health, and safety and operations. Usually the management representing each area would be the department head or a designated representative of the department head.

Top-down commitment from the management present in the operation is required for a management-of-change process to be successful. There must be swift and harsh ramifications taken toward those members of the group who shortcut the system for nothing more than convenience or flexibility. Likewise, groups that utilize the system effectively should be granted recognition for embracing the concept and leading by example. The underlying structure of the management-of-change process should be detailed in a formal policy with relevant procedures detailing how to use the system. The best management-of-change policies will be written in a manner that provides provisions for emergency situations. These emergency excursions usually can only be granted for initial approval by a select handful of highly qualified personnel. It should require more than one individual's approval, even in an emergency situation, to ensure that sufficient review and thought can be generated before granting or denying the request. Once an emergency change request is granted, however, the change would still be reviewed by the entire committee to ensure appropriateness at a later date. Most well-written policies will note that the product being manufactured under the emergency situation will be placed under a precautionary hold status (no shipment) until the broader group of experts have deemed the change appropriate. Building this type of flexibility into the change process still allows for the business to be supported at times when key personnel may be absent, yet provides rigor around the serious nature of making changes in a formal, dispassionate manner.

Formal signatures should be required as part of the documentation of the change, and the original request should be in written format even if it is simply documentation of a verbal conversation (a telephone call at 2:00 a.m., for example, or a voicemail message). By documenting each request in a traceable manner, the program can be audited to ensure its ongoing success and compliance of the processes operating in its revised mode. A typical management-of-change request form would include: (1) The name of the originator, (2) the date the request is made, (3) the desired effective date of the change, (4) the current operational settings, (5) the proposed operational settings, and (6) a brief description of why the change is desired.

The key for ensuring that the management-of-change procedure is successful lies in the vigilance of the personnel responsible for managing it. The amount of flexibility allowed for emergency situations is critical as well. The management-of-change system should provide a balance between unnecessary bureaucracy and common sense so that business needs may be met without putting the brand at risk. To ensure that every change made is not an "emergency," the people who have the authority to grant emergency approvals should be aggressive and stringent in their limited review period. By dealing with emergency approvals in this manner, others will view the process as structured and thorough versus a necessary rubber stamp method. Performing random audits of both emergency and planned change requests that were documented will ensure that the system remains functional over time. As is true of any policy or procedure, however, initial and ongoing training must be provided to the people in the facility who will be affected (supervision, maintenance, operators, auditors, managers, etc.).

8.6 Planning for equipment failure and other emergency situations

A typical worst-case equipment failure, when customer-service needs are extremely high, often occurs on either a weekend, or off shift. These times usually correlate to periods where the personnel most qualified to make decisions or repair the equipment are not present. Knowing that adverse events will happen, the astute processor will plan ahead and develop policies and procedures that detail acceptable practices for resuming operations. The procedures should detail the location of spare parts/equipment, how to repair or replace various equipment, and how to conduct a mini-validation/verification study to ensure that the equipment functions as designed once the repairs are complete. The procedures must make reference to a documented policy that contains clear guidance for when production may resume in the event of inspection equipment failure. Sometimes repair of equipment may not be possible, and production without the detection equipment in place is ruled to be an unacceptable risk that cannot be assumed.

Unfortunately for some unknown reason, when inspection equipment fails in the absence of clear policy, the operation will usually employ some additional visual inspections and continue running until instructed to do otherwise. A great amount of suspect product can be manufactured over an 8 to 12-hour period in a food manufacturing facility. An emergency contact list provided to the plant personnel is definitely warranted to ensure that poor decision-making processes do not jeopardize business results. In some instances the processor may want to initiate a rotating on-call list that defines finite resource availability so extensive downtime periods are avoided. It is extremely important to realize that these events can happen in any manufacturing facility; without proper proactive planning expect to personally face the downside risk of such poor decisions.

Works Cited and Select Resources

1. *Procedures for Performing a Failure Mode, Effects and Criticality Analysis.* 1980. Military Standard 1629A. Washington, D.C.: U.S. Dept. of Defense. 54 p.

2. Stevenson KE, Bernard DT. 1995. *HACCP—Establishing Hazard Analysis Critical Control Point Programs. 2nd ed.* Washington, D.C.: The Food Processors Institute. 224 p.

3. Lock AP. 2002. *Metal Detection.* In: Food Plant Sanitation. Pederson, N.Y.: Marcel Dekker Inc. 752 p.

4. Neal C. 2003. *Prerequisites for Successful Validation.* J of Validation Technology 9(3):240–245.

5. Snyder JE. 2002. *Corrective and Preventative Action: Planning to Achieve Sustainable GMP Compliance.* J of GXP Compliance 6(3):29–39.

CHAPTER 9

Proper Use of Acceptance Sampling and Statistical Process Control to Augment FM Control Programs

9.1 Introduction

When working to prevent foreign material contamination of products there comes a time when even the best intended hypothetical conjecture, expert risk assessment, and established past history reach their limitations. No written book, audit event, or government inspection can effectively monitor the variation in a supply chain . . . that is a task that only the processor can assume responsibility for. Certainly monitoring product characteristics with tangible data is necessary when crossing over from the dimension of theoretical food safety/quality to the dimension of real world food manufacturing.

What data are you currently collecting about your raw materials or processes that provides assurance that your product is free of physical hazards or otherwise in an optimal state of quality when it leaves the facility? Could you stand in front of a regulatory agency or customer and prove that proper due diligence was taken while Product X was being manufactured? Do you know how many samples are necessary to disposition lots with statistical confidence on a routine basis? What kind of sampling do you need to perform if you suspect that a lot is nonconforming in respect to foreign material? How do you set up challenge studies for new detection or separation equipment that is placed in your facility? Simply stated, it is impossible to substantiate fact-based conclusions to yourself and others without possessing appropriate data and documentation. This chapter was written to provide you with proper conceptual guidance for going forth and collecting data necessary to display conclusions with statistical confidence. Many individuals fear statistics and mathematical concepts, but to properly protect your brand and the public health it is imperative that a quantitative approach to the disposition of food products is embraced. A simple gut feel or group nod of acceptance usually does not provide legal standing for processors or individuals who are implicated in litigation. This chapter is lengthy and comprehensive to ensure that the reader is well versed in these important concepts.

The tenets of applied statistics can be your strongest allies when trying to comprehend the ever-present variability of modern complex production processes. The various statistical methods outlined in this chapter can be utilized by the reader to analyze process data and better quantify the existing risk to brand equity. Two essential tools that any professional in the food processing industry should understand are acceptance sampling and statistical process control (SPC). SPC and acceptance sampling are two overlooked candidates for preventing foreign material contamination of foods. In most instances, use of data derived from acceptance sampling, process studies, and SPC monitoring are the only acceptable sources of scientific evidence to justify the monitoring frequency and critical limits of HACCP plans.

Throughout the chapter we will explore the application of statistically valid sampling strategies and examine their utility in foreign material control programs. Remember that extensive sampling protocols alone are not the only mechanism that should be used to prevent defective units from entering the food supply. Additionally, the use of acceptance sampling to disposition lots of finished product suspected of containing foreign materials of public health significance (as defined in Chapter 1) should never be considered as appropriate due diligence by a processor.

A comprehensive foreign material control program rooted in science-based tools such as HACCP, and strict adherence to GMPs throughout manufacturing, storage, and distribution is the current best approach. Quality and safety must be built into the product as it is manufactured; no other system will provide adequate protection to the consumer or the producer over the long term. For this reason a two-tiered approach to sampling intermediate and finished products while monitoring process controls and sanitary conditions is necessary for achieving success. Remember, using a foreign material control system that relies entirely on finished product inspection can become quite costly to the producer upon the first lot rejection or recall.

9.2 Types of data that can be gathered during inspection or process monitoring

The process of sampling products or monitoring process outputs can generate two types of data for analysis depending on the collection strategy. The first data type is known as "attribute data," and is sometimes referred to as "categorical data." Attribute data usually represents the output of a categorization process that can only possess a distinct outcome. This outcome is inherently mutually exclusive of all other outcomes for the process/event (3). We can use a baseball analogy to better illustrate the underlying concept of attribute data. When a pitcher throws a baseball toward an eager batter, the batter will either swing the bat or let the pitch go by. If the pitch is caught by the catcher, it is then determined by the umpire to be either a strike or a ball. The umpire does not have the luxury of making up a new category and, once the pitch is categorized, the process is over. Likewise, if the batter swings at a pitch, he or she will either hit the ball or miss it. Note that in the examples above there are only two distinct, mutually exclusive outcomes used to characterize the events. If data were collected on these specific components of a baseball game, it would be attribute data.

An example of attribute data generated in a manufacturing environment can be found when a producer is inspecting product for shipment. The producer either accepts or rejects a product, based on certain characteristics, and the product will either be shipped to the trade or placed on hold. Notice that all of the results presented could be coded into a two-digit system of 0 = pass and 1 = fail (following a binomial pattern). Once the data has been numerically coded, it can be analyzed with statistics using any number of theoretical distributions.

Yet another example of using attribute data can be found when one attempts to trend consumer complaints received by a company for alleged foreign material. Most companies create distinct categories for their consumer complaints to ensure that the data is actionable for those charged with driving continuous improvement. Possessing knowledge that a company has received 50 consumer complaints for foreign material per million units shipped doesn't help the team that is trying to resolve the underlying root cause. If the complaints were placed into subcategories by a knowledgeable individual and graphically represented, as in Figure 9.1, it is easy to visualize where effort needs to be placed. This type of categorical analysis is a powerful way to utilize attribute data. Later in the chapter we will discuss trending techniques that will allow you to use SPC to trend attribute data. For further reference several examples of attribute data collection can be found in Table 9.1.

The second type of data that can be collected and subsequently analyzed is variable data. Variable data is defined as data that, when collected, is expressed within any continuous interval numerically (i.e., a whole

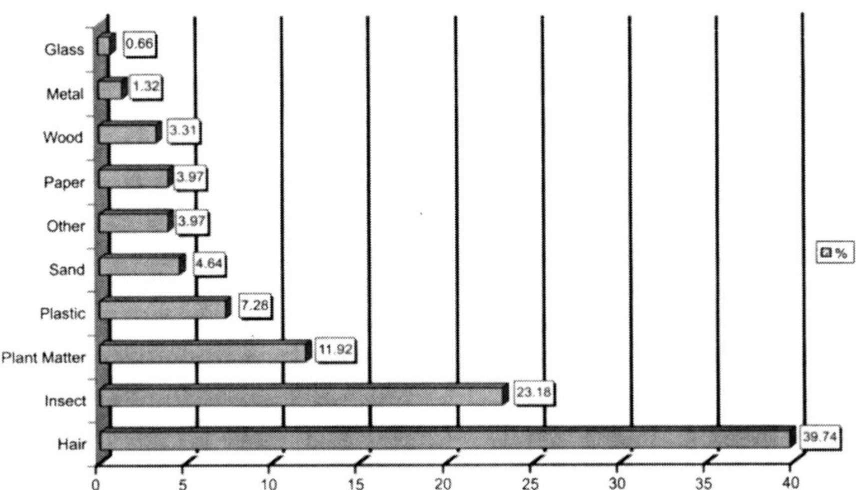

Figure 9.1 Chart of overall complaints represented in a pareto format for ease of comprehension.

Table 9.1 Common examples of attribute data collection in the food industry.

Description of Data Collection Activity	*Typical Result*
Number of metal detector kickouts in a shift	18
Number of container defects found in a lot of bottles	60
Count of consumer complaints received in a period of time	500 per month
Number of units held for off quality versus shipped to the trade	800/12,000
Count of insect fragments found in a flour sample	32

Table 9.2 Common examples of variable data collection in the food industry.

Description of data collection activity	*Typical Result*
Net content determination of a package	100.5 g
Measuring the length and width of diced product	25 cm × 8 cm
Determining the viscosity of a product	1200 cp
Recording the temperature of a freezer	15°C
Measuring the pH of a liquid product	4.3

number, fractional component, negative value, or decimal). Going back to the baseball example discussed earlier, a variable measurement could have been made on the speed of the pitch (90.4 miles per hour), or the overall distance that the ball traveled (260 feet) after it was hit by the batter. A great manufacturing example of variable data can be found when examining particle size distributions when performing a sieve analysis. Typically, the components of the dry blend will be categorized by their size via physical transport of the material through a series of progressively smaller mesh sieves. A certain quantity of material is retained on each sieve because it is physically too large to pass through. The material left on each sieve is usually then weighed, and a numerical value representing the subcomponent weight (100.59 grams) is recorded for the sieve. Other examples of variable data collection activities are illustrated in Table 9.2. Variable data, like attribute data, can also be readily analyzed with statistics, but for the sake of brevity we will exclude variable acceptance sampling and variable SPC from this text.

9.3 Statistically valid sampling plans versus tribal practices

Now that the concept of data collection types has been fully developed, the concept of acceptance sampling will be presented. Acceptance sampling has been used by industry for decades to decide the fate of various products, and can be performed using either attribute or variable data. Most acceptance sampling activities involving foreign material in the food industry, however, deal with attribute data as defects are classified and enumerated. Acceptance sampling is the process by which a subset of a production lot is sampled and evaluated against a standard to determine whether the entire lot conforms to the established specification(s) (2). This end-of-the-line approach to quality is utilized by firms to disposition products (i.e., hold or ship) where a 100 percent inspection routine is impractical. Such situations of impracticality are usually driven by the cost-prohibitive nature of the inspection activity itself, or by the fact that the inspection requires functional testing that is destructive in nature (i.e., ensuring that a round manufactured for a military rifle does, in fact, fire).

The food industry is filled with sampling plans and systems that were "validated" through years of use. The intricacy of organizations' quality systems that originated from such tribal roots can be incredible. Policies, procedures, and work instructions are usually generated, lost, and generated again from the experience and lore captured by previous generations of former members. When one digs deeper at the historical roots of many of these documents, it can be a very enlightening experience. Granted that a company is only as good as its institutional knowledge base, many companies continue to operate under false assumptions of their risk management structure. By failing to adopt sampling plans based on statistical merit, as opposed to tribal methods, companies may continue to operate at an undetermined level of risk.

Statistically valid sampling strategies are one facet of quality systems that many companies find challenging to address properly. The examples presented in this chapter will not reveal the identity of the producers implicated or go into specifics of where poor sampling plans were witnessed. These examples are meant, however, to be specific conceptual models of what should, and sometimes should not, be done to protect brand equity and public health. The typical tribal sampling plans are present in many forms, but realize that none of them really help you quantify the true risk of failure that your company or customer/consumer may be burdened with. Some typical plans that are lacking statistical merit are outlined below—please do not panic if your company's plan is similar to those outlined in the following section because valid sampling strategies are introduced later in the chapter to aid you.

The first sampling strategy to be evaluated was debunked by Juran, and involves the notion that sampling ten percent of the units in a production lot is sufficient to deem the lot "free" of defects (2). Hopefully the notion of sampling one in ten containers when seeking defect levels of less than 1 per 1,000 units seems ridiculous and not worthy of consideration for protecting your brand equity. Unfortunately, this strategy has been used in produce grading applications and purity analyses of meat products by major producers across the United States. Figure 9.2 represents the adversity a sampling plan will face in trying to identify a 1 in 1,000 defect scenario.

The prospect of destructively testing ten percent of each lot produced is also not an appealing ongoing business strategy for the producer. Later in the chapter, when the operational characteristics (OC) curve is developed, it will be easy to contrast the protection offered by this plan with those of more contemporary statistically valid strategies.

Yet another common strategy of simply sampling the square root of the number of units within a lot is lacking in statistical merit as well. The origins of this method are uncertain, but the applications have been broad. In fact the FDA alluded to this type of sampling methodology in their

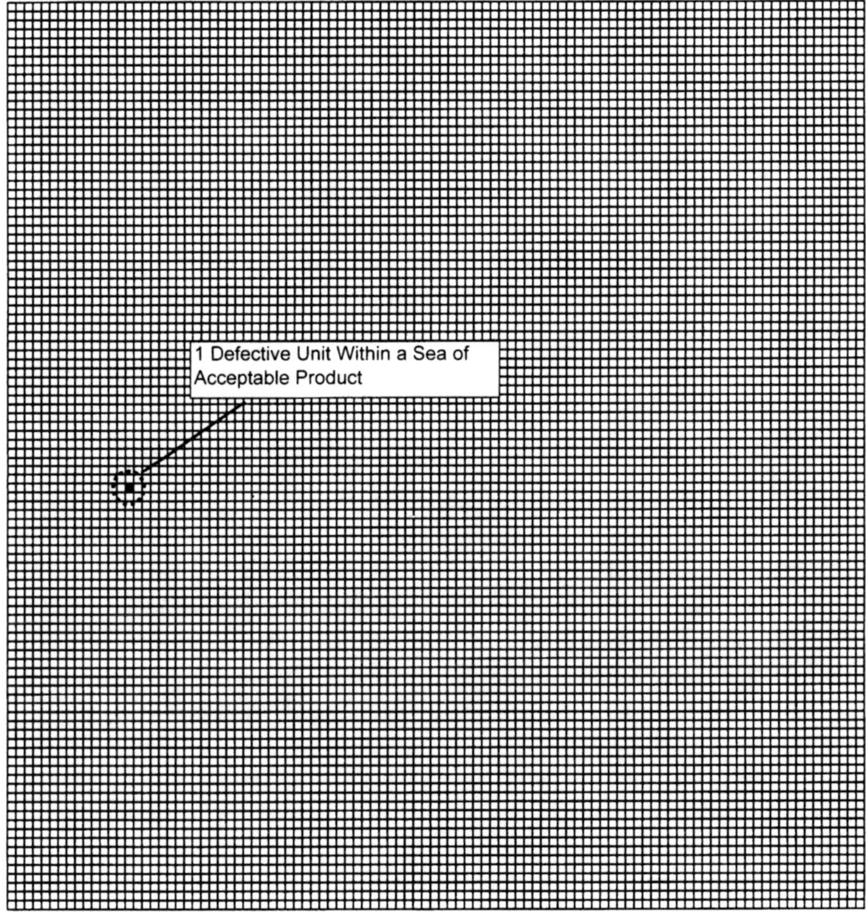

Figure 9.2 Graphical representation of the defect Level 1 unit per thousand units. The single black square represents the defective unit.

Investigators Operations Manual as recently as the 1995 edition. The idea itself seems logical, and simple to implement. Just envision an employee on a shipping dock entering the number of total units into a pocket calculator and hitting the square root key—poof, a sampling plan is born. The simplicity seems quite endearing, but unfortunately the plan is not without fallacies as will later be apparent with the introduction of the OC curve.

A final acceptance sampling pitfall is to use a transient internal or external expert to set up a statistically valid sampling plan. The initial approach is valid, but if the sampling plan is never updated as the process continues to evolve, a serious failure may occur. Many producers have, in fact, preserved the same sampling strategies that were in use decades ago for lot acceptance criteria, only to have their lot sizes grow by tenfold or more as they leverage greater economies of scale. Producers will also misapply the definition of a lot to include product from multiple production lines over different days in an attempt to reduce inspection costs. Whatever protection was presented by inspecting 50 cases at the end of a 1,000 case run surely will not be the same as a 50-case sample from a 10,000 case run mixed lot scenario. An example like this clearly shows that the intent of the original objective has not been met, unless the only goal of the sampling plan was to ensure the destruction of 50 cases for accounting purposes.

9.4 What constitutes a statistically valid acceptance sampling plan?

So what is a producer to do, knowing that the last plans detailed are not truly valid in a statistical sense? Obviously, sampling for the sake of sampling is not an effective strategy for a quality assurance program. Inspection of product components in the food industry is usually an expensive and tedious exercise that is performed by humans of varying levels of ability (minus the automated processes discussed in earlier segments). Therefore, sampling strategies should only be used where the risk of failure justifies expenditure of resources and a 100 percent inspection strategy isn't practical or necessary.

It is extremely important for you, the reader, to grasp the concept that acceptance sampling is all about identification and management of foreseen risk. If we didn't know what type of defects could occur, we probably wouldn't attempt to put a sampling strategy together that identifies the defect in a given lot of raw material or finished goods. As a producer it is critical to identify what the objective of the sampling plan is going to be before expending resources on sampling products. Likewise, actionable procedures for handling the results of the inspection need to be outlined prior to commencing inspection. The first question you must ask of your company before drafting inspection protocols is what level of outgoing quality is acceptable for your brand for a given defect type? This level of risk is known as the specification acceptable quality limit (Spec-AQL) (3). For many senior managers this is a very difficult question to address and, when faced with having to make the decision, many senior managers will simply state, "Zero." The concept of zero risk is of noble merit, but in actuality it is impossible to guarantee a zero consumer/customer risk position for your company, unless you cease to manufacture and distribute products.

From a risk management and financial perspective, allowing differing levels of allowable defects per type of FM is quite appropriate. It is probably a safe assumption that an excess amount of chaff would more likely be better tolerated in an incoming load of wheat flour than aluminum shavings from beverage containers. Most producers break their product specifications into multiple classifications of defects, giving each a maximum defect level. The general norm for classifying defect types is by grouping them into critical, major, or minor classifications where each is assigned a corresponding Spec-AQL. Using the example of the wheat flour above, the producer might have treated the excess chaff as a minor defect, whereas the metal contamination could be classified as a critical defect. Clear objectives on what the plan is to deliver for assurance must be set up in advance of blindly sampling a lot, or interpreting results from a sample set that had previously been taken. Typical statements of objective can be as simple as "to ensure to a 95 percent confidence interval that defect X is less than three occurrences per 1,000 opportunities in lot Y."

The second step to adopting an acceptance sampling strategy is to actually demonstrate that the sampling plan can deliver a result that meets your overall objective. Previous historical data, current process data, and controlled studies are needed to meet this objective (2, 6). Some strategies used by processors can be as simple as 100 percent inspection of several lots produced under the new acceptance sampling planning and documentation of the results. Other strategies utilized are to seed known "good" lots with known quantities of defective products to identify the efficacy of the inspection method and sampling plan. Obviously, before any of these strategies are used, the inspection method must be formalized and those who are to implement it should be thoroughly educated. A common pitfall for many processors is to employ an inspection strategy that is highly people-dependant, put forth a tremendous amount of effort to validate the approach, and then place untrained employees in the inspection role. Obviously, no quantity of upfront effort expended in planning and validation can remedy that kind of situation.

9.4.1 Readily available acceptance sampling plans for use

Currently there are many published sampling plans that have been used across industries as a common practice for acceptance sampling of attribute and variable data. The first plan to discuss is the Mil-Std-105 series. The 105 series was published by the Department of Defense as a procedure for inspection of products by attributes to be used by contract manufacturers producing goods for the military. The Mil-Std-105-E (4) was the most recent and final revision of the series; the final publication of Mil-Std-105-E occurred in February 1995. Many producers still use the Mil-Std-105-E as a best practice for acceptance sampling even though the Department of Defense no longer maintains the standard. A second widely used sampling plan for attribute data is maintained by the American National Standards Institute (ANSI) in collaboration with the American Society for Quality (ASQ). The ANSI/ASQ Z1.4 sampling plans for attributes takes up where Mil-Std-105-E left off, but differs in some components of definition and sampling rules (5). Both Mil-Std-105-E and ANSI/ASQ Z1.4 provide tables of AQLs and sampling frequencies to be used for given lot sizes and differing levels of scrutiny. Acceptance plan switching rules for reduced inspection quantities are also presented in the ANSI/ASQ Z1.4 plan, and they do differ from those in Mil-Std-105-E. A third widely used attribute acceptance sampling plan is published by the International Organization for Standardization (ISO). ISO 2859-1 is published and maintained for professionals to use; however, for the sake of brevity we will not delve into comparing or contrasting its components to the previously identified plans within this text.

9.4.2 Properly using sampling plans to discriminate between the good, the bad, and the ugly—an overview of Mil-Std-105-E

It should be obvious by now that acceptance sampling is definitely a science in itself, and we need to ensure that we do not "art" it up. By performing acceptance sampling properly, you can quantify your risk and save precious resources versus 100 percent inspection. Remember, though, that the act of sampling product alone does not guarantee that it is free from foreign material. Acceptance sampling must be utilized in a calculated manner to accent established quality assurance and food safety programs . . . NOT as a direct substitute.

So how does one use the published standards properly to make good risk determinations? For the purposes of this section we will use Mil-Std-105-E as our tool of choice. Don't fret, because the use of 105-E is almost identical to that of the ANSI/ASQ Z1.4, so once one is mastered the others should come quite naturally. This will become more apparent once the common underlying statistical basis for the plans has been reviewed. Before we go into how to use the Mil-Std for acceptance sampling, we must first explain the components of this plan so that informed decisions may be made.

The first 16 pages of the Standard are filled with introductions and definitions of the terms used in the plan. There is a decent set of instructions on how to use the standard presented within the plan, but presentation of more enhanced instruction here will ensure that you are able to use the standard properly. Table 1 (reproduced here as Table 9.3) is the first interface that the plan presents for the user to begin. There are four basic decisions you must make at Table 1 (Table 9.3) herein before reaching a functional sampling plan:

1. The AQL that you want to begin with
2. The lot size of your inspection
3. The inspection level you wish to use (I, II, or III)
4. The sampling scenario you wish to use (single, double, or multiple)

As for the AQL definition, there are 26 levels to choose from within the Mil-Std, from 0.010 to 1,000. The definition of the AQL for the purposes of the Mil-Std is different from the concept of the specification AQL defined earlier. In the Mil-Std scenario, the AQL is the established level of quality that you would not want to routinely reject at a high frequency as an ongoing process average. Therefore, when using Mil-Std-105-E, understand that the AQL you choose is that which has a high probability (usually greater than 87 percent) of acceptance if the product is as good or better than the AQL. A common misconception by novice users of the Mil-Std is that the AQL chosen represents the maximum defect level of a lot that has a high probability of being rejected before it can potentially reach the consumer/customer. Unfortunately, this is not the case due to a fundamental concept of statistics, known as error, that is confusing to most

Table 9.3 Table I adopted from Mil-Std-105-E.

TABLE I – **Sample Size code letters**

(See 4.9.1 and 4.9.2)

Lot or batch size			Special inspection levels				**General inspection levels**		
			S-1	S-2	S-3	S-4	I	II	III
2	To	8	A	A	A	A	A	A	B
9	To	15	A	A	A	A	A	B	C
16	To	25	A	A	B	B	B	C	D
26	To	50	A	B	B	C	C	D	E
51	To	90	B	B	C	D	D	F	G
91	To	150	B	B	C	D	D	F	G
151	To	280	B	C	D	E	E	G	H
281	To	500	B	C	D	E	F	H	J
501	To	1200	C	C	E	F	G	J	H
1201	To	3200	C	D	E	G	H	K	L
3201	To	10000	C	D	F	G	J	L	M
10001	To	35000	C	D	F	H	K	M	N
35001	To	150000	D	E	G	J	L	N	P
150001	To	500000	D	E	G	J	M	P	Q
500000	And	Over	D	E	H	K	N	Q	R

CODE LETTERS

people. Let's back up and look at the underlying hypothesis, or assumptions, of why a producer would test a portion of a lot versus the entire lot.

The underlying fundamental principle of the Mil-Std-105-E and ANSI/ASQ Z1.4 is that the lot is good and should be accepted, but we need to test samples of it to prove so. This principle is grounded in an ethical assumption that if a producer were to knowingly make a bad lot, they would not ship it (at least we all hope not). Obviously the other outcome of the inspection activity referenced above is that, if the inspection results cannot support that the lot is good, it should then be rejected. It is important to remember that, when your overall decision-making process is not based on definitively examining 100 percent of the items in a lot, you are making an educated assumption with the probability of being incorrect.

Fortunately for us these probabilities can be calculated using statistical inference. As with any base assumption there is a calculated probability that you can be wrong on two counts (See Figure 9.3):

1. You could reject your assumption when it is actually correct (this is known as a Type I, or α error) (3). In this case you wouldn't accept the lot even though it met your established AQL. This is also referred to as the "producer's risk," because doing this routinely could result in a severe financial burden for the company.
2. You could accept your assumption when it is actually incorrect (this is known as a Type II, or β error) (3). This is also referred to as the "consumer/customer's risk," because routine shipment of lots like this would not be tolerated well by your consumer/customer.

The reason that this is the case is because standards like Mil-Std-105-E and ANSI/ASQ Z1.4 were written with the basic hypothesis of protecting the producer from scrapping lots possessing quality that is as good or better than the AQL. The flip side of this statement is that the sampling plans presented in these standards are not lacking in power to definitively say a lot has a high probability of routinely being rejected at calculated levels above the AQL. These levels of limiting quality are referenced in the Standard's various tables. We will discuss how to use tables within each standard to understand what the consumer/customer risk is for each plan,

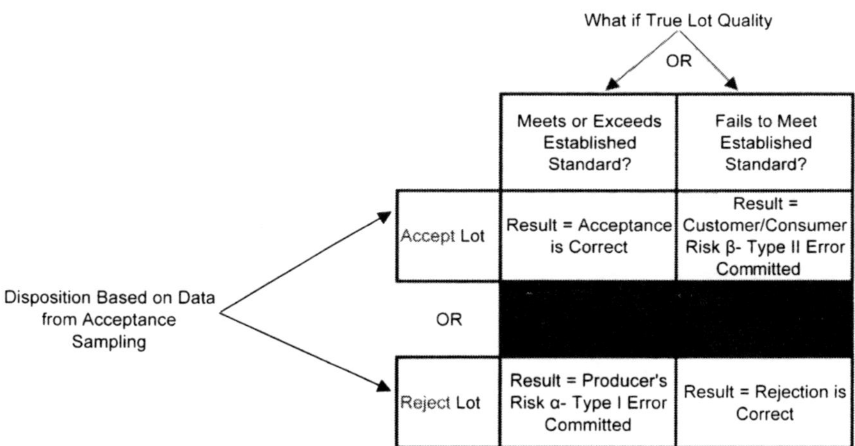

What if True Lot Quality
OR
Meets or Exceeds Established Standard?
Fails to Meet Established Standard?
Disposition Based on Data from Acceptance Sampling
Accept Lot
OR
Reject Lot
Result = Acceptance is Correct
Result = Customer/Consumer Risk β- Type II Error Committed
Result = Producer's Risk α- Type I Error Committed
Result = Rejection is Correct

Figure 9.3 Graphical representation of Type I and Type II errors committed when basing lot acceptance on statistical sampling versus the results of 100% inspection.

but in the meantime remember the reference to AQL in Table 9.3. is really a number of acceptance that is based on limiting the producer's financial risk of scrap product, not consumer/customer protection.

The next nugget of information you need to use for the sampling tables is the lot size that you are attempting to inspect. This is a pretty straightforward question and refers to the collection of units of product from which a sample is to be drawn and inspected. As an illustrative example, let's look at a soup manufacturer's plight to disposition a lot of product that is on hold for various defects.

9.4.3 Reactive Soup LLC—a case study in the use of Mil-Std-105-E

Imagine yourself as the QA Manager for Reactive Soup LLC, which produces soup in three-liter retortable pouches for foodservice and institutional use. One of the main particulate ingredients in your best selling Vegan Delight soup is frozen green beans diced to approximately 4 cm in length. Your sole bean supplier, Stalkville Farms, has had a difficult crop year from a quality standpoint, and has not been able to supply enough beans for you to meet customer demand for Vegan Delight orders. To make matters worse, Reactive Soup has just been issued formal notification from Stalkville that one of their rejected lots was inadvertently shipped to your facility. Due to a failure in their sorting system, the rejected lot contained an excessive level of attached stems and extraneous fibrous plant material. Reactive's purchasing specification for attached stems in the frozen bean dices is 4.5 percent by count, and less than one percent extraneous plant material by weight. From the limited sampling that Stalkville has done, you are able to ascertain that the rejected lot contained approximately 11 percent attached stem defects and 4 percent extraneous plant material. Thankfully, you were quickly able to reduce Reactive's risk exposure to an isolated 9,000 pouch run through review of the bean usage records over the past week of production.

The Vice President of Reactive's Institutional Products Division does not want to scrap this lot because the production management believes that their sorting inspectors performed exceptionally well on the day in question in spite of the challenges present. Vegan Delight is definitely a hot seller, and past history records show that Reactive has occasionally shipped lots that are at a one percent overall prevalence level (equal to one defective pouch per 100) of defectives for stem and woody material contamination, with few reported issues. In Reactive's case a defective pouch of soup is defined when there is any combination of > two attached stems and/or > one piece of visible fibrous plant material per pouch. Reactive's history illustrates that complaints are more likely when defects exist greater than this established level. You must now develop a sampling plan to address Reactive's predicament and determine the risk of shipping the soup to the customer/consumer. Fortunately for you the last QA manager left a copy of the Mil-Std-105-E in the desk drawer, so your task has become slightly less stressful.

If you have 9,000 institutional-size pouches of soup that must be inspected to ensure that a one percent defect level of pouches would not be rejected or thrown away at a regular frequency when it is actually good, your lot size would simply be 9,000. Your first point of reference when using the Mil-Std will be looking at Table 1 (Table 9.3). Simply use the lot or batch size from Table 1, where the interval contains the lot size in question. In this instance you would see that 9,000 units is actually within the lot or batch size interval of 3,201 to 10,000, so you would use that row of the table to find the corresponding sampling table code letter.

The next step to complete in order to identify a sampling plan that can be used from the Mil-Std is to identify what inspection level you would like to follow. Essentially, if you have never performed sampling of your product and have limited experience with a supplier/processor, poor confidence in a supplier/processor, or limited data to base a decision on, use of inspection Level III (also referred to as tightened inspection) is recommended as a starting point. This level of inspection offers the most discrimination for avoiding both Type I and Type II errors. In our example of the potentially "good" soup, we will use inspection Level II (normal) as the row to identify which sampling plan code letter we will refer to. Note that in this instance the code letter "L" is identified for our use. Within both Mil-Std-105-E and ANSI/ASQ Z1.4, there are rules presented by which one could utilize reduced inspection if need be, but in this case reduced inspection is obviously not advisable. The ANSI/ASQ rules do differ from those of the Mil-Std, so be sure to reference each accordingly before adopting an ongoing strategy of reducing inspection for your business.

The final step in choosing the sampling plan is to decide whether a single, double, or multiple sampling strategy is appropriate for your situation. For this example a single sampling plan will be used. One could use a double or multiple acceptance sampling strategy and achieve the same protection, but in this case the single sampling plan is a wise choice. We must now look to the single sampling plan tables for normal inspection in the subsequent pages that use code letter "L" for our AQL of one percent. Table II-A (reproduced as Table 9.4) displays the proper table that should be referred to in this instance—note that the sample size for code letter "L" is 200 pouches with a lot acceptance of up to five defective pouches found in the course of inspection. If six pouches were found in the course of inspection, you can feel comfortable in rejecting the lot based on the high probability that the actual defect level is not equal to or less than the one percent producer's risk that you were willing to assume.

Now that we have covered the producer's risk for the sampling strategy outlined above, we need to understand the consumer/customer risk associated with the sampling scheme. Ideally one would not set up a sampling strategy without fully understanding the consumer/customer risk associated with it. In the case of the one percent code "L" plan that we adopted in the previous example, the 90 percent confidence interval of consumer risk can be found by examining Table VI-A within the Standard. Table VI-A (reproduced as Table 9.5) illustrates the limiting quality (LQ) of a corresponding sampling plan. When we look to the LQ (also referenced as Long Term Projected Defect Level) table, we find that for sample code "L" at a Level II (normal) inspection there is a 90 percent probability of not accepting a lot containing 4.6 percent defects. At first glance this may seem astonishing, but realize that this LQ/LTPD represents your risk of assuming a Type II error. If a 95 percent assurance of LQ is desired by the processor, we can determine it by using Table VII-A (reproduced as Table 9.6). Put yourself in the role of the QA Manager for Reactive Soup LLC (which produces this soup), who determines that an inspected lot containing a defect level of 5.3 percent would likely not be accepted. Note that when a higher level of assurance is desired, a higher potential defect level is listed in the table. For some this concept is difficult to grasp, so analysis of the OC curve for this sampling plan is appropriate.

No matter which plan you choose to utilize, one basic component that underlies the risk assessment in acceptance sampling remains the same—that of the OC curve. The OC curve for a sampling plan is the tool that helps you quantify the risk associated with using a specific acceptance sampling protocol. An OC curve for an attribute plan is simply a graph of the fraction defective within a lot versus the probability that the specified sampling plan will accept the lot. Remember that no sampling plan will provide you with absolute risk assurance, but at least examining an OC curve from an established attribute sampling plan can give you a quantitative estimation of the risk.

Table 9.4 Table II-A adopted from Mil-Std-105-E.

TABLE II-A – Single sampling plans for normal inspection (Master table)

See 4.9.3. and 4.9.4)

Sample size Code Letters	Sample size	Acceptable Quality Levels (Normal Inspection) 0.010	0.015	0.025	0.040	0.065	0.10	0.15	0.25	0.40	0.65	1.0	1.5	2.5	4.0	6.5	10	15	25	40	65	100	150	250	400	650	1000
		Ac Re	Ac Re	Ac Re	Ac Re	Ac Re	Ac Re	Ac Re	Ac Re	Ac Re	Ac Re	Ac Re	Ac Re	Ac Re	Ac Re	Ac Re	Ac Re	Ac Re	Ac Re	Ac Re	Ac Re	Ac Re	Ac Re	Ac Re	Ac Re	Ac Re	Ac Re
A	2	↓	↓	↓	↓	↓	↓	↓	↓	↓	↓	↓	↓	↓	↓	0 1	↓	↓	1 2	2 3	3 4	5 6	7 8	10 11	14 15	21 22	30 31
B	3	↓	↓	↓	↓	↓	↓	↓	↓	↓	↓	↓	↓	↓	0 1	↑	↓	1 2	2 3	3 4	5 6	7 8	10 11	14 15	21 22	30 31	44 65
C	5	↓	↓	↓	↓	↓	↓	↓	↓	↓	↓	↓	↓	0 1	↑	↓	1 2	2 3	3 4	5 6	7 8	10 11	14 15	21 22	30 31	44 65	↑
D	8	↓	↓	↓	↓	↓	↓	↓	↓	↓	↓	↓	0 1	↑	↓	1 2	2 3	3 4	5 6	7 8	10 11	14 15	21 22	30 31	44 65	↑	↑
E	13	↓	↓	↓	↓	↓	↓	↓	↓	↓	↓	0 1	↑	↓	1 2	2 3	3 4	5 6	7 8	10 11	14 15	21 22	30 31	44 65	↑	↑	↑
F	20	↓	↓	↓	↓	↓	↓	↓	↓	↓	0 1	↑	↓	1 2	2 3	3 4	5 6	7 8	10 11	14 15	21 22	↑	↑	↑	↑	↑	↑
G	32	↓	↓	↓	↓	↓	↓	↓	↓	0 1	↑	↓	1 2	2 3	3 4	5 6	7 8	10 11	14 15	21 22	↑	↑	↑	↑	↑	↑	↑
H	50	↓	↓	↓	↓	↓	↓	↓	0 1	↑	↓	1 2	2 3	3 4	5 6	7 8	10 11	14 15	21 22	↑	↑	↑	↑	↑	↑	↑	↑
J	80	↓	↓	↓	↓	↓	↓	0 1	↑	↓	1 2	2 3	3 4	5 6	7 8	10 11	14 15	21 22	↑	↑	↑	↑	↑	↑	↑	↑	↑
K	125	↓	↓	↓	↓	↓	0 1	↑	↓	1 2	2 3	3 4	5 6	7 8	10 11	14 15	21 22	↑	↑	↑	↑	↑	↑	↑	↑	↑	↑
L	200	↓	↓	↓	↓	0 1	↑	↓	1 2	2 3	3 4	5 6	7 8	10 11	14 15	21 22	↑	↑	↑	↑	↑	↑	↑	↑	↑	↑	↑
M	315	↓	↓	↓	0 1	↑	↓	1 2	2 3	3 4	5 6	7 8	10 11	14 15	21 22	↑	↑	↑	↑	↑	↑	↑	↑	↑	↑	↑	↑
N	500	↓	↓	0 1	↑	↓	1 2	2 3	3 4	5 6	7 8	10 11	14 15	21 22	↑	↑	↑	↑	↑	↑	↑	↑	↑	↑	↑	↑	↑
P	800	↓	0 1	↑	↓	1 2	2 3	3 4	5 6	7 8	10 11	14 15	21 22	↑	↑	↑	↑	↑	↑	↑	↑	↑	↑	↑	↑	↑	↑
Q	1250	0 1	↑	↓	1 2	2 3	3 4	5 6	7 8	10 11	14 15	21 22	↑	↑	↑	↑	↑	↑	↑	↑	↑	↑	↑	↑	↑	↑	↑
R	2000	↑	↑	1 2	2 3	3 4	5 6	7 8	10 11	14 15	21 22	↑	↑	↑	↑	↑	↑	↑	↑	↑	↑	↑	↑	↑	↑	↑	↑

⇓ = Use first sampling plan below arrow. If sample size equals or exceeds lot or batch size, do 100 percent inspections

⇑ = Use first sampling plan above arrow.

Ac = Acceptance number

Re = Rejection number

SINGLE NORMAL

Table 9.5 Table VI-A adopted from Mil-Std-105-E.

TABLE VI-A – Limiting Quality (in percent defective) for which P_a = 10 Percent
(for Normal Inspection, Single sampling)

(See 3.3)

Code Letter	Sample Size	Acceptable Quality Level															
		0.010	0.015	0.025	0.040	0.065	0.10	0.15	0.25	0.40	0.65	1.0	1.5	2.5	4.0	6.5	10
		Ac Re	Ac Re	Ac Re	Ac Re	Ac Re	Ac Re	Ac Re	Ac Re	Ac Re	Ac Re	Ac Re	Ac Re	Ac Re	Ac Re	Ac Re	Ac Re
A	2															68	
B	3														54		
C	5													37			58
D	8												25			41	54
E	13											16			27	36	44
F	20										11			18	25	30	42
G	32									6.9			12	16	20	27	34
H	50								4.5			7.6	10	13	18	22	29
J	80							2.8			4.8	6.5	8.2	11	14	19	24
K	125						1.8			3.1	4.3	5.4	7.4	9.4	12	16	23
L	200					1.2			2.0	2.7	3.3	4.6	5.9	7.7	10	14	
M	315				0.73			1.2	1.7	2.1	2.9	3.7	4.9	6.4	9.0		
N	500			0.46			0.78	1.1	1.3	1.9	2.4	3.1	4.0	5.6			
P	800		0.29			0.49	0.67	0.84	1.2	1.5	1.9	2.5	3.5				
Q	1250	0.18			0.31	0.43	0.53	0.74	0.94	1.2	1.6	2.3					
R	2000			0.20	0.27	0.33	0.46	0.59	0.77	1.0	1.4						

LQ (DEFECTIVES)
10.0%

Table 9.6 Table VII-A adopted from Mil-Std-105-E.

TABLE VII-A – Limiting Quality (in percent defective) for which $P_a = 5$ Percent
(for Normal Inspection, Single sampling)

(See 4.11)

Code Letter	Sample Size	Acceptable Quality Level															
		0.010	0.015	0.025	0.040	0.065	0.10	0.15	0.25	0.40	0.65	1.0	1.5	2.5	4.0	6.5	10
		Ac Re	Ac Re	Ac Re	Ac Re	Ac Re	Ac Re	Ac Re	Ac Re	Ac Re	Ac Re	Ac Re	Ac Re	Ac Re	Ac Re	Ac Re	Ac Re
A	2															78	
B	3														63		
C	5													45			66
D	8												31			47	60
E	13											21			32	41	50
F	20										14			22	28	34	46
G	32									8.9			14	18	23	30	37
H	50								5.8			9.1	12	15	20	25	32
J	80							3.7			5.8	7.7	9.4	13	16	20	26
K	125						2.4			3.8	5.0	6.2	8.4	11	14	18	24
L	200					1.5			2.4	3.2	3.9	5.3	6.6	8.5	11	15	
M	315				0.95			1.5	2.0	2.5	3.3	4.2	5.4	7.0	9.6		
N	500			0.60			0.95	1.3	1.6	2.1	2.6	3.4	4.4	6.1			
P	800		0.38			0.59	0.79	0.97	1.3	1.6	2.1	2.7	3.8				
Q	1250	0.24			0.38	0.50	0.62	0.84	1.1	1.4	1.8	2.4					
R	2000			0.24	0.32	0.39	0.53	0.66	0.85	1.1	1.5						

LQ (DEFECTIVES)
5.0%

Figure 9.4 shows the OC curve for sample plan "L" utilized in the soup example. Interpretation of the graph is quite intuitive when the intercept for the AQL of one percent is illustrated as a reference point. Note that when you read the OC curve at the AQL of 1.0 percent, it is at the 95 percent probability level of acceptance. At this level there is a five percent chance of making a Type I (α) error (rejecting the lot when it meets the AQL). By following the X axis from the AQL intercept over to the origin of the Y axis, it is easy to visualize the protection that is offered to the producer when using Mil-Std-105-E. The slope of the line rises to the 99th percentile within a 0.31 percent defect level. Now let's look at Figure 9.5, which displays the LQ/LTPD 90 as well. By examining the graph it is easy to visualize the apparent customer/consumer risk presented by the sampling activity ending in a Type II (β) error. This risk is visualized by examining the numerical distance across the X axis before you reach the ten percent probability of acceptance. Following this logic, at even the 3.0 percent defect level (three times the producer's risk), there is almost a 50 percent probability the lot would be accepted. Figure 9.6 and Table 9.6 illustrate the protection gap between the AQL of 1.0 percent and the LTPD with a five percent chance of acceptance. Figure 9.7 displays the OC curve

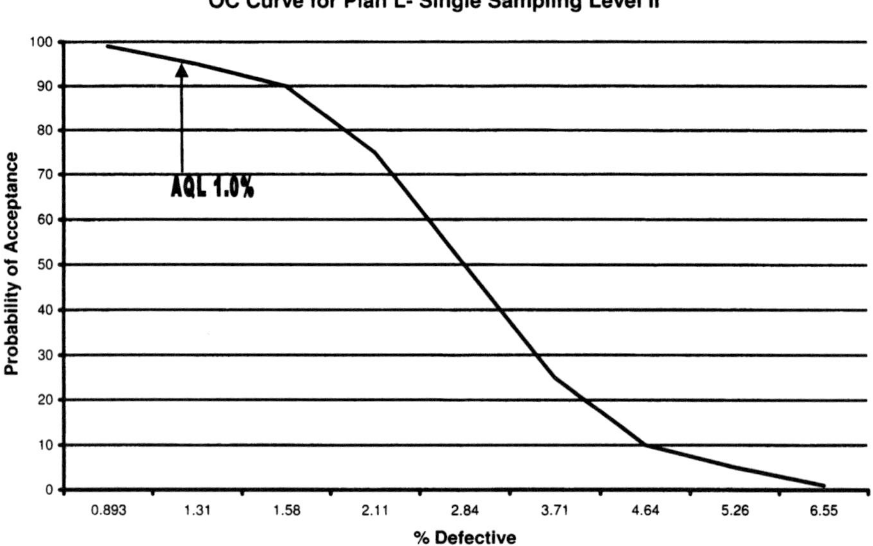

Figure 9.4. OC Curve for Plan L adapted from Mil-Std-105-E.

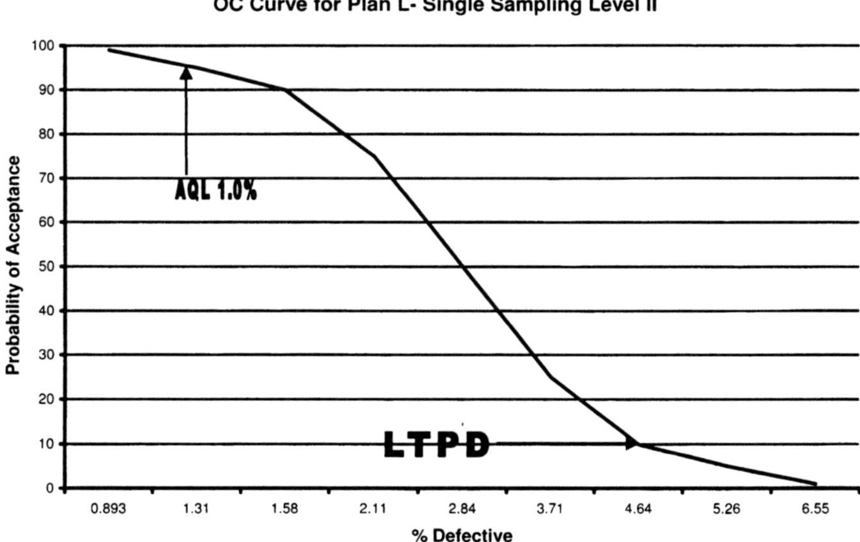

Figure 9.5. OC Curve for Plan L displaying the LTPD at a 10% probability of lot acceptance.

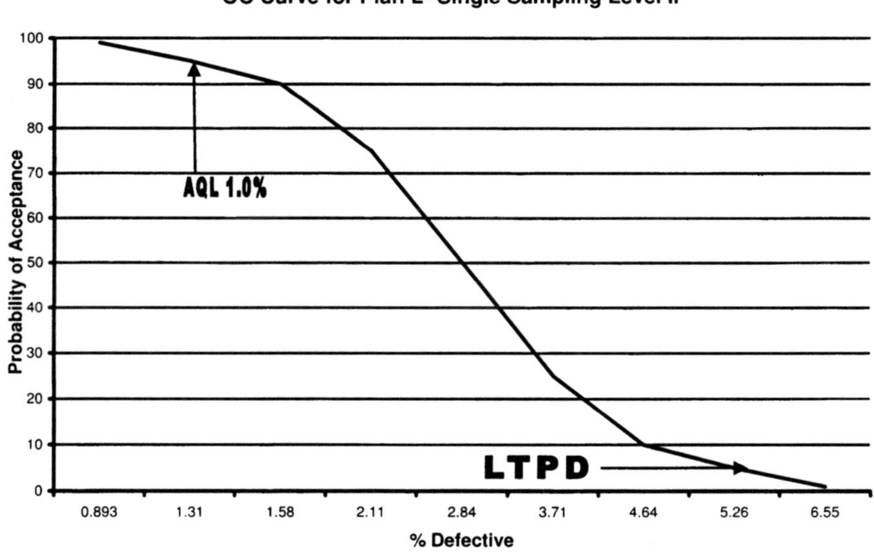

Figure 9.6. OC Curve for Plan L displaying the LTPD at a 5% probability of lot acceptance.

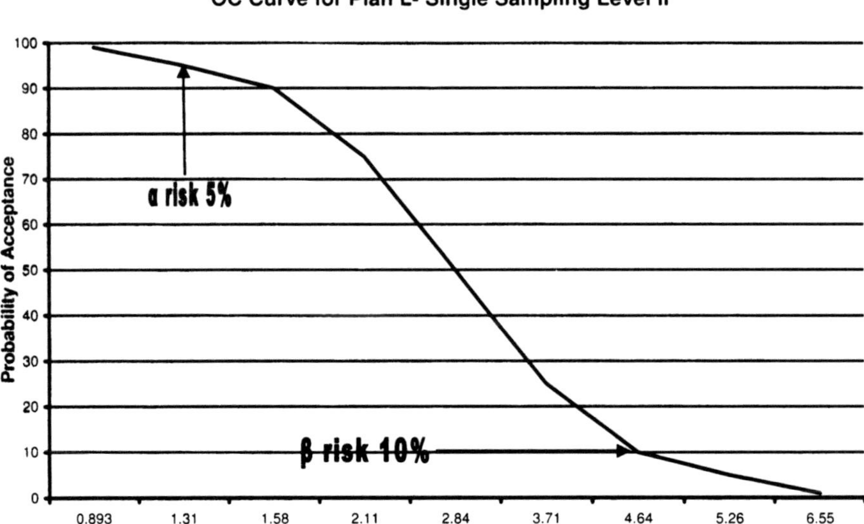

Figure 9.7. OC Curve displaying the producer's risk—α, and the customer/consumer risk—β present when an LTPD of 10% acceptance is used.

again, this time illustrating the α and β risk margin associated with the sampling plan.

In Reactive's case the sampling plan is implemented the next day by employees trained in identifying the defects, and the magic number of six defective pouches was met within the first 100 samples inspected. The sampling activity ceased as soon as the sampling tolerance had been reached and the inspection results were documented. After looking at the inspection results and the graphs of the potential LTPD, the vice president requested that the lot be donated to a local charity and a large financial claim pursued against Stalkville.

9.5 Always remember the β, and other acceptance sampling pitfalls

Hopefully the soup example helped display the critical linkages between applications of acceptance sampling and dealing with FM issues. The goal of the example was to illustrate that, when acceptance sampling plans are used properly, they can help protect the producer and the consumer from undue risk. The point must be repeated that, when using published sampling plans, it is crucial that the lot inspection strategy be based on both the AQL and the LQ/LTPD. Unfortunately, many producers make the classic mistake of basing sampling strategies solely on AQLs, only to be bitten by the β risk of Type II error. Another common error that must be avoided is improper definition of the true lot. Lots should be defined based on the homogeneity of the product that was produced, and not be composed of multiple production dates lumped together for convenient sampling and subsequent disposition.

Typical results of failing to avoid Type II errors and improper lot definition can include increased consumer complaints, loss of customers, product seizures, litigation expenses, and rejection of product at receiving docks located thousands of miles away from the manufacturing facility. It is critical, when documenting procedures that reference use of acceptance sampling, that the processor must also document the AQL and LQ/LTPD of the sampling plan chosen. By doing this in a standardized manner, it will be much easier to present the risk associated with sampling activities to senior management, legal counsel, and/or regulators at a later date if necessary.

9.6 Is it truly random sampling?

Any result of a statistical sampling strategy is only as good as the randomization that took place prior to drawing the sample. One of the classic examples that comes to mind is when producers go through the trouble of carefully choosing a sample size, then carefully discern the lots to be inspected, only to hand the task of gathering the samples over to people with no instruction on how to procure them. Some even try to justify that this is a random process in itself, as in "How do you know which boxes they'll choose . . . it seems pretty random to me." From a statistical sense, however, the premise of randomness revolves around the premise of equality; that is, for any random sampling technique to be statistically valid all of the units within the sampled lot must have an equal opportunity of being selected for inspection.

With respect to some of the tribal methods that I have witnessed, it is conclusive that cases of product at the bottom of pallets are hardly afforded equal opportunity for sampling by uninstructed personnel. The most extreme case of tribal sampling mentality I have ever witnessed involved whole vegetable products that were being held because of suspected contamination with human hair. The producer had rationalized that, by randomly choosing the samples from a single pallet, all the necessary random samples could be obtained to disposition the lot. The problem with this approach was that there was no data or documentation to attest to the fact that the problem would have been more prevalent at any time frame in the 17-hour period of sequential production, which encompassed 60 pallets. Surely there are better ways to ensure that sampling patterns are truly random in nature than the pluck and pray method described above.

The first step to ensuring randomness, therefore, is to set up a systematic numbering scheme that can identify all possible samples within a lot. For example, imagine a producer having to conduct an inspection on a lot of 240 drums of juice concentrate spread evenly across 60 pallets (4 drums per pallet) for the presence of undesirable bentonite contamination. The producer needs to sample 20 drums for sensory and microphysical analysis prior to receipts being issued to the supplier. A typical tribal method would be to instruct the technician to choose a drum off of every third pallet to take samples from and get them back to the laboratory quickly. The proper technique to assure randomness in this case, however, would be to assign each drum within the lot a number. Once each drum was assigned a numerical identity, a lottery could be conducted using a statistically fair method to identify which 20 drums would be sampled. Going through the trouble of first assigning a numerical system to the lot ensures each drum a 1/240 chance of being the first sampled. Each drum chosen in the following sample iteration would have a 1/240-N probability of being chosen (where N is the number of sample trials previously taken). The fairness of the tribal method previously described is suspect because the technician may favor the drum on each pallet that is easiest to access, possesses the most legible label, or is on the "good luck" corner of his/her favorite pallet.

Thankfully there are many mechanical and virtual methods for ensuring that a statistically fair lottery is conducted. When dealing with smaller sample sizes, certified casino dice work quite well for identifying units for sampling. For intermediate sample sizes, a (fair) bingo drum can do the trick effectively, and can make the process fun for the technicians as well —just imagine the joy of attempting to justify the off-site trips or invoices that will be necessary to procure such items to the members of the Accounting Department.

A less labor-intensive method of identifying samples, but slightly boring, is to use a table of random number seeds readily available in most statistical reference books. A more robust solution, however, is to utilize a random-number generator housed within a software program. Microsoft's Excel has a random number generation function that is functional and easy for technicians of varying aptitudes with computers to use. Once boundaries are established via user input, a simple copy and paste function of the formula can generate as many random numbers as needed for the would-be sampler. The only drawback to the computer approach of generating a sampling strategy is that in FDA-regulated environments ambiguity around 21CFR Part 11 may open the producer up to some extra due diligence on the validation of the electronic sampling plans generated.

9.7 Using SPC to monitor the prevalence of potential foreign material contaminants

Acceptance sampling will always be necessary in some capacity within the food industry, but it can co-exist quite well with the application of statistical process control (SPC) schemes. The topic of SPC in this chapter will be limited for the sake of brevity to the discussion of attribute data. Certainly variable data collection offers more power in terms of potential analyses, but in order to develop the topic fully another book of this length would need to be dedicated solely to the subject.

Rather than continuously utilizing acceptance sampling to disposition lots that are potentially contaminated with foreign material, the producer is better served by applying the same effort in developing proactive FM trending plans based on SPC principles. The amount of inspection that must be applied to a lot to assure an LTPD of 0.115 percent defectives using an acceptance sampling strategy is 2,000 samples with a lot acceptance of zero defects and a lot rejection of one. In contrast, using SPC proactively could spread out this sampling over a longer production period to monitor various raw materials or equipment. This approach would further aid the processor to predict and prevent failure versus constantly reacting to incidences of FM contamination.

Making the leap to successfully utilizing attribute SPC for FM control is not much of a stretch for most processors. In most cases food processors are already collecting data in attribute format, whether they recognize it as such or not. This activity is usually performed by personnel who are responsible for monitoring process controls, sanitation, or HACCP on paper forms assembled as part of the production record. The company's purpose for recording such routine monitoring activity is usually to maintain compliance with regulatory requirements or internal policies. Converting to attribute SPC from this point is relatively simple, seeing as the data collection activity is already a standard practice. By quickly implementing attribute SPC, the producer usually will gain an entirely new level of risk assessment/management for the minor costs associated with training and generation of new forms for producing SPC charts. SPC software packages of varying cost and complexity, such as the Synergy series by Zontec, are also readily available for purchase as a turnkey solution.

In truth the most difficult aspect of managing the transition to SPC isn't generating forms or buying software, but rather breaking down old cultural paradigms associated with the monitoring function. It can be quite difficult to influence long-time employees to act on statistical trends within the data versus comparing each measurement to a standard and then filing the results away forever. Like all successful change strategies, the cultural keys to successful SPC implementation in such an environment are:

1. Provision of a high-quality effective conceptual education to the employees affected prior to execution. This training should not be based on the premise of turning average employees into statisticians.
2. The new data-dependent statistical thinking culture must be modeled by the leadership group at all times.
3. Consistent hands-on mentoring of employees using the tool is necessary until they repeatedly display mastery of the principles.
4. Provision of positive reinforcement for those who are correctly using the system.
5. A function must take ownership for auditing and maintaining the SPC systems over time (i.e., calculating control limits, including new processes in the scope, summarizing overall results, etc.) for long-term success.

One fundamental aspect of a databased SPC culture is the recognition that all processes and their subcomponent material streams possess inherent variability. To comprehend the scope and severity of this variability, one must be able to properly choose the information to be collected. Only with the collection of relevant information can the variability be interpreted into succinct but meaningful statements of fact. For this reason machine output and material characteristics are best understood when represented in a numerical format and interpreted using mathematics. Looping back to the Reactive Soup example, presented previously, allows us to analyze a case where SPC could and should be effectively utilized to monitor the process and control FM.

9.8 Reactive gets proactive

Obviously destructively testing products of institutional-sized portions that cannot be readily reworked is a cost-prohibitive proposition for any processor. Reactive Soup learned an expensive lesson when the company rejected the lot of soup (9,000 pouches) for high defects, and then had to short shipments to one of its key customers. Enlightened by this experience, the management of Reactive now would like to implement novel preventative measures to ensure that another supplier failure will not cost them future business results. In order to aid Reactive, the bean preparation/handling process must be mapped and in-process sampling points identified to implement SPC (see Figure 9.8).

The best spot to successfully trend the variation in the bean material stream was found to be immediately after the inspection belts. This step of Reactive's process is just prior to where beans are sent to the batching area via large transfer bins. Notice that quality trending for the beans is performed prior to the beans being commingled with other ingredients. If a failure is discovered at this step of the process, corrective action can be applied and any beans that needed to be further sorted or scrapped would not be blended with other ingredients or packaging materials. This approach provides protection for the producer in that the cost of poor bean quality will be restricted to the bean ingredient processing operation, and may not require expensive finished product to be scrapped. An additional advantage of placing material evaluation at this stage of the process is that it affords the producer a data stream to evaluate the sorting operation. The trending of bean quality over time should allow labor costs associated with the sorting operation to be better managed. If the incoming quality of beans gradually improves, the data may illustrate that a reduction in sorting labor could be achieved without sacrificing the ingredient quality. Conversely, the data might also identify additional sorting that staff will be required, in the bean-sorting operation, to achieve parity quality levels over difficult periods.

Let's now re-examine Reactive's manufacturing process with the application of SPC at the green bean sorting station. A trained operator is now positioned at the end of the sorting station, where an automated

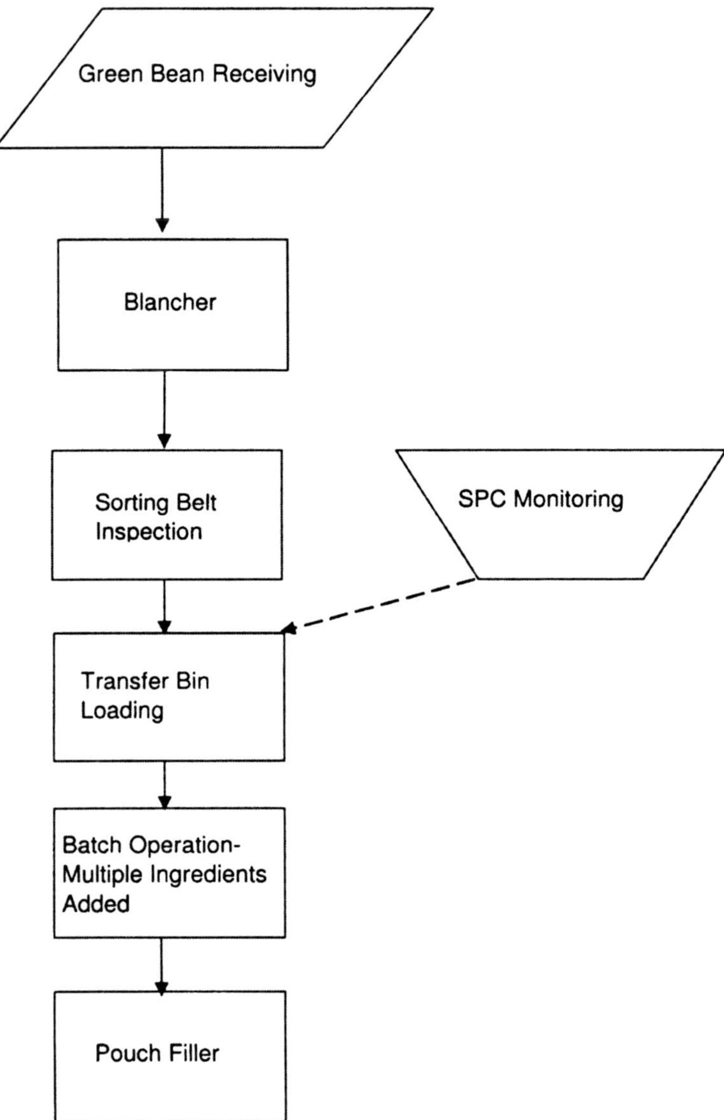

Figure 9.8. Process diagram of the green bean ingredient flow at Reactive Soup LLC.

pusher bar diverts a full cross section of the inspection belt four times within each hour of production (this operation yields approximately one sample every 15 minutes). Each sample generated for analysis is composed of approximately 0.8 kg of green bean dices. The operator stationed after the sorting station will count the number of dices present, then sort out any defects by category. The work instruction for the area details that any sample reaching a five percent defect level is unacceptable, and the operator must ensure that the beans are not sent on for further processing. The instructions also clearly state that any bins filled since the last good check should be withheld from batching when a five percent defect level is present. As a measure of prevention, any sample analysis resulting in greater than a 3.5 percent defect level will require the operator to reduce the speed of the sorting belt. If this corrective action is not proven to be effective by the data gathered at the next check, additional inspectors are to be procured by the area supervisor.

Once the operator finishes enumeration and classification of the defects, the results are entered into a computer log sheet (Table Tabular Inspection Results), and a P chart for attribute trending of the FM defects is generated and displayed in ready view of all the employees in the sorting and transfer area. A P chart was chosen for this application because it can be used for trending percent defectives in cases where the occurrence of a defect is not a rare event (could be greater than 5 percent). The computer calculated the upper control limit (UCL) of 4.71 percent by using three standard deviations (see formula below), and by using data obtained from samples taken over the previous crop year. Note that the UCL in this case is not an arbitrary number, but is based on previous performance data and is calculated using the formula (6)

$$UCLp = \mathrm{P}\backslash + 3 \times \sqrt{(P\backslash \times (1 - P\backslash)) \div n}$$

where

$$P\backslash = (Total\ \#\ Defectives) \div (Total\ \#\ Inspected).$$

The computer also calculated the process average of 1.4 percent based on previous performance data. With the process behavior characterized in a graphical manner, the operator can draw conclusions from data very quickly versus using gut feel or trying to interpret cryptic tabular data. The work instruction for the area explicitly states that if a run rule is violated or a result falls outside the upper control limit, the operator is to take necessary corrective action to bring the process back under control.

Operators can react quickly and uniformly to trends in foreign material defects when they are properly trained to interpret control charts like the P chart example above. Many food processing operations still rely on collecting tabular data for quality assurance and HACCP monitoring. In this scenario most operators fail to react to trends, and typically the only action taken is when an obvious out-of-specification condition presents itself. Properly governed operations using control charts are usually much more sensitive to process changes and provide better proactive troubleshooting than such traditional methods. The P chart for the first day of running is displayed in Figure 9.9. Note for Day 1 that 31 samples were tested by the operator, and the process operated in line with its previous history. The conclusion that the sorted bean quality is likely no different than that encountered in previous history is supported by the fact that the chart does not display out-of-control points. Further evidence of an in-control

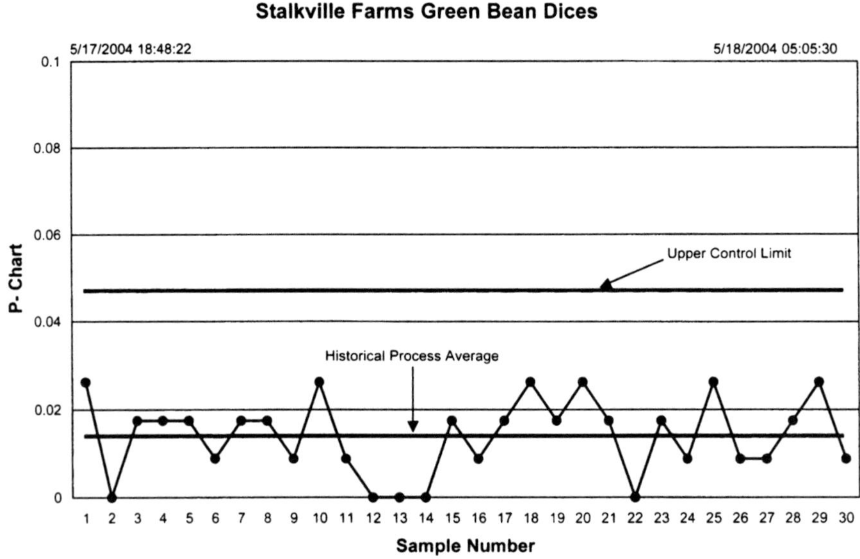

Figure 9.9 P-chart generated from the first day of green bean sorting data obtained at Reactive.

process is present in the observation of a random distribution of points around the process average. Additionally none of the typical process shift "run rules," as characterized by W.A. Shewhart in 1931, were violated in this graphic.

A review of the second day of production reveals some interesting patterns. By reviewing Figure 9.10, it is apparent that the quality of produce has shifted. Note the extensive sawtooth pattern at Sample 40 followed by Samples 50 through 59—all being greater than the process average. The statistical probability of having nine sequential samples yield results greater than your historical process average and still being able to conclude that the process has not shifted (i.e., does not contain more defects) is extremely remote. At this point we could begin to hypothesize that either the sorting operation is ineffective or incoming produce quality has deteriorated significantly. Note in this instance that a process shift was detected by the P chart *before* a specification exceedance was ever documented. Would the same corrective action from trending result if the operator's only interface from the inspection activity was tabular data (presented in

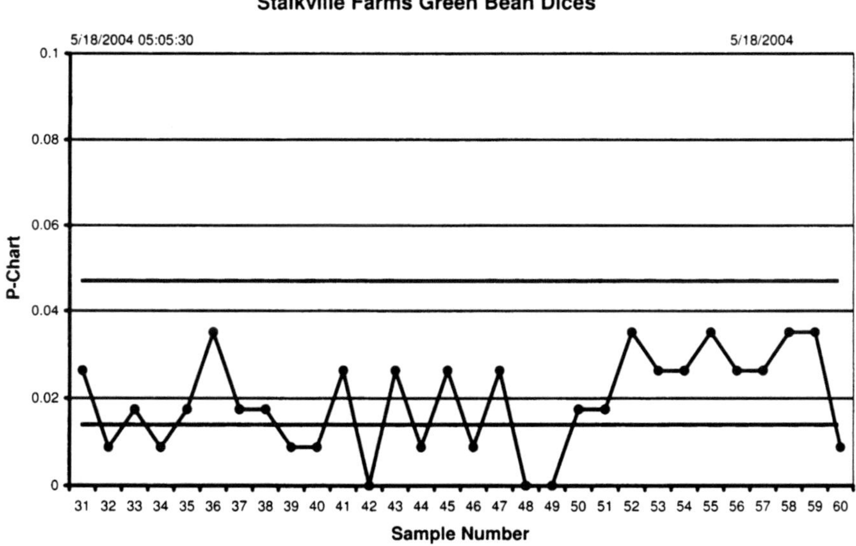

Figure 9.10 P-chart generated from the second day of green bean sorting data obtained at Reactive.

Table 9.7). It would be prudent for Reactive to investigate the potential root cause of the process shift prior to continuing production. A subtle shift of ingredient quality would likely not have been noticed in Reactive's previous mode of operation, and a corrective action would not have been taken.

The P chart for the third day of production is represented by Figure 9.11. The last sample for the production day displays a sample that exceeds the upper control limits, and is clearly out of specification. Obviously, the ingredient quality and/or sorting is not up to par with its past history. Reactive makes the decision to shut down production of Vegan Delight at this point to further investigate the situation. If Reactive had stuck with their old operational strategy (no SPC) and no corrective action was taken, a bad situation may have unfolded.

Let's look at what the next two production runs would have netted for defects if Reactive were to have continued uninterrupted operation of the green bean sorting area without SPC. Figure 9.12 illustrates the increase in FM defect levels over time that would correlate to a shift in green bean quality due to a process failure at Stalkville Farms. If green beans from the last day of sorting were sent on to the soup-batching operation, surely the finished product defect level would not have been met. This could have led to another potential costly scenario for Reactive.

Remember that when applying attribute SPC in an operation to prevent FM contamination, there is still a risk of committing Type I or Type II errors with each sample taken. Thankfully an OC curve can be generated to display the producer and customer/consumer risk associated with the sampling activity. If the sorting process were to generate 4.3 kg of green bean dices per minute to support the soup-batching operation, approximately 64 kg of green beans would have been generated within any 15-minute period of production (assuming 100 percent reliability of the sorting process). Specifically in the case of attached stems, we are interested in the proportion of defects by count. Recording the overall number of beans sampled is important in determining the power of the SPC monitoring. In Reactive's case the number of bean dices enumerated per sample does vary slightly, but consistently averages 315 dices per 0.8 kg sample (sorting through this quantity of beans requires a lot of activity by the operator within a 15-minute period). The 64 kg of beans that pass by the station equals approximately 25,400 bean dices. Even with the sample size and lot size varying over time, the resultant OC curve for the sampling plan (presented in Figure 9.13) illustrates the excellent protection offered for Reactive.

Table 9.7 Representation of the Reactive Soup monitoring process. Note the amount of difficulty that an operator would face in spotting a statistical trend within the sea of numbers.

Sample #	*Defects*	*Sample* #	*Defects*	*Sample* #	*Defects*	*Sample* #	*Defects*	*Sample* #	*Defects*	*Sample* #	*Defects*
1	3	21	2	41	3	61	1	81	1	101	7
2	0	22	0	42	0	62	2	82	2	102	6
3	2	23	2	43	3	63	3	83	3	103	8
4	2	24	1	44	1	64	4	84	1	104	6
5	2	25	3	45	3	65	4	85	3	105	0
6	1	26	1	46	1	66	4	86	3	106	4
7	2	27	1	47	3	67	0	87	5	107	1
8	2	28	2	48	0	68	2	88	4	108	7
9	1	29	3	49	0	69	2	89	1	109	5
10	3	30	1	50	2	70	3	90	1	110	7
11	1	31	3	51	2	71	1	91	6	111	2
12	0	32	1	52	4	72	1	92	4	112	6
13	0	33	2	53	3	73	1	93	7	113	8
14	0	34	1	54	3	74	1	94	7	114	2
15	2	35	2	55	4	75	4	95	7	115	2
16	1	36	4	56	3	76	2	96	4	116	6
17	2	37	2	57	3	77	4	97	6	117	4
18	3	38	2	58	4	78	0	98	6	118	7
19	2	39	1	59	4	79	4	99	3	119	2
20	3	40	1	60	1	80	4	100	6	120	3

*Average of 3 defects per inspection

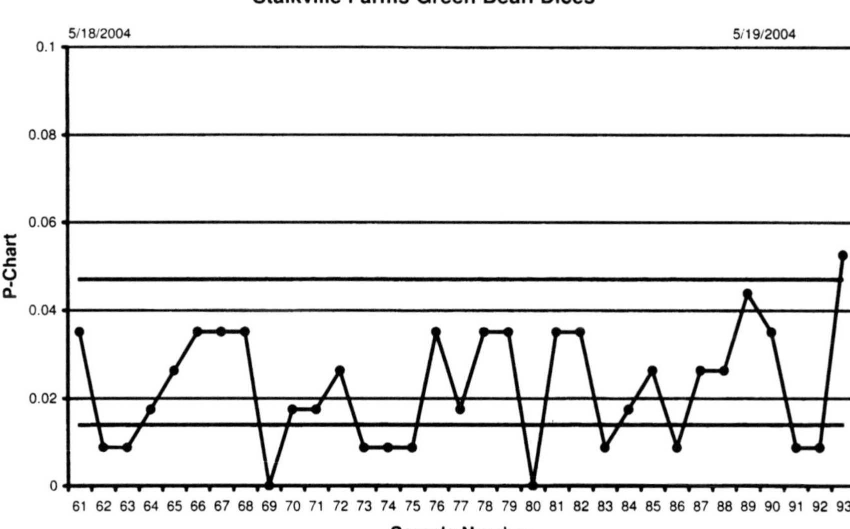

Figure 9.11 P-chart generated from the third day of green bean sorting data obtained at Reactive.

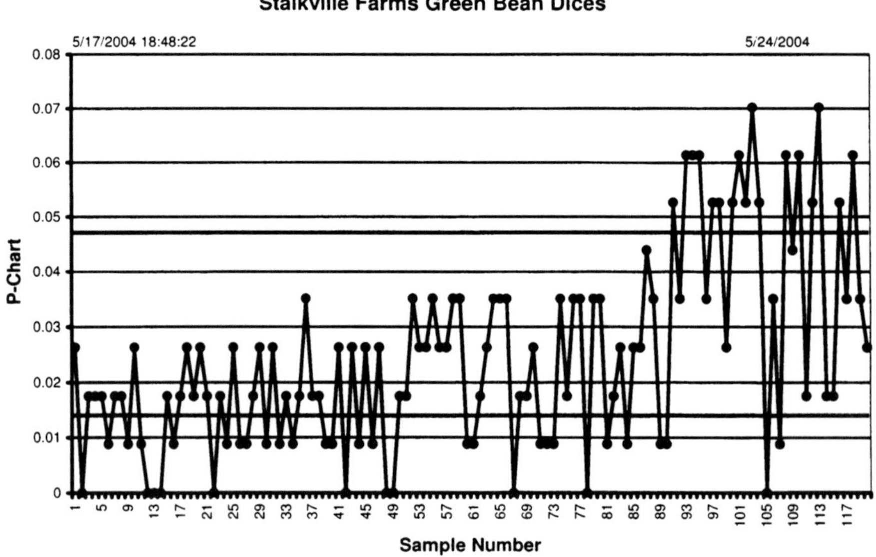

Figure 9.12 P-chart assembled to represent what a failure at Stalkville Farms would provide as observed quality at the Reactive sorting station.

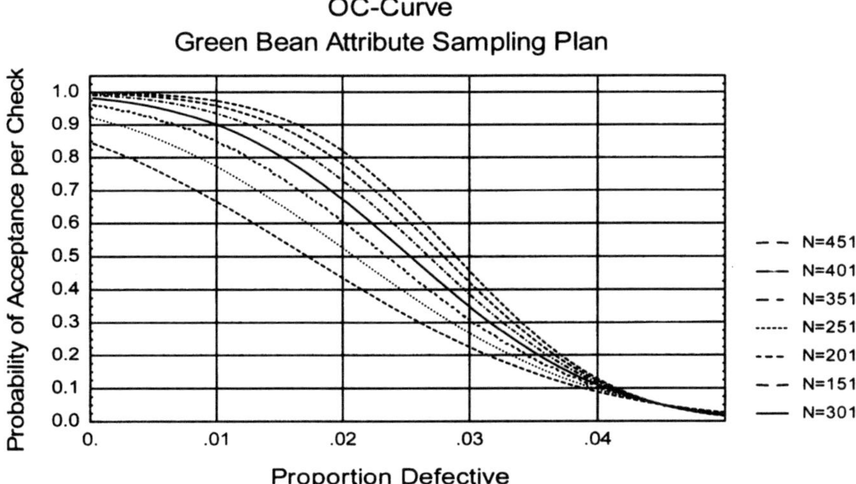

Figure 9.13 OC Curve for the attribute SPC sampling being conducted at Reactive.

9.9 Other examples of applying attribute SPC to help prevent foreign material contamination of foods

The Reactive examples presented earlier in the chapter were meant to be instructive in nature. Hopefully the advantages of using SPC are now apparent, and the overall concept is well understood. Understand that there are a plethora of uses for SPC as a means of preventing FM in the food industry. The balance of this chapter is dedicated to presenting examples that can be applied quickly to shore up common loopholes in foreign material control programs.

9.9.1 Tracking finds on magnets

Documentation of the corrective actions for finds on magnetic separators is always a wise policy for a processor. A more prudent strategy is to trend finds using attribute SPC charts. Typically magnets are placed in areas, as outlined in previous chapters, where there is a potential for magnetic FM contaminants. Most processors collect data in tabular format on finds and review the results against internal standards for the purposes of disposition. When data is collected from magnets that are screening raw materials, the quality of those materials can be trended over time using categorical data analysis. The advantage here is that trends can be spotted and addressed prior to major contamination incidences unfolding.

Using SPC charts for magnet finds of raw ingredient streams is an excellent application. Diligent monitoring of the process can afford a healthier purchaser/supplier relationship when working in a mode of continuous improvement. Data-based communication can usually occur between the two parties before lot rejections or other potential riskier scenarios unfold.

Without SPC data the conversations between suppliers and processor are usually filled with anecdotal facts that don't offer much potential for investigation. Typical comments such as, "We usually don't have a lot of trace finds, but something seems different lately . . . " are many times not actionable for the supplier to investigate. Possessing data that is relevant to specific supplier lots used within dates of operation that are suspect for magnetic FM provides a much better basis for the supplier to begin investigation.

Using SPC to trend magnet finds in intermediate processing points is also a value-added activity. All equipment wears and can break down over a period of use. Influences such as operational settings, lack of preventative maintenance, or substitution of original equipment manufacturer (OEM) parts with cheaper alternatives can dramatically alter the mean time between such failures. Failure modes that could generate magnetic FM should mostly be identified based on a hazard analysis of the process. By trending magnet finds in a categorical manner, the processor can use the data to better anticipate mass equipment failure that could lead to a catastrophic product event.

Figure 9.14 represents in-process magnet finds from a supplier of organic flour to the food industry. When examining the chart it was not difficult to notice routine spikes in trace-metal finds over time. The root cause of the contamination was determined to be the result of a maintenance activity on a screener performed by a transient, inexperienced technician. When working in the area the technician would increase the hertz on the drive for the vibratory separator and tighten the tolerance between the racks that hold the screens. He did this to better prevent the separator from plugging (a major driver of downtime for the processor) so he could be free to perform other activities besides laboriously cleaning the separator. Unfortunately the oscillation of the unit was enough to compromise the newly adjusted clearances, and metal-to-metal contact was created. Thankfully, the vibration had caused the screen racks to separate back to a greater clearance and the problem would dissipate as the operation continued. By reviewing the data in the form of a control chart, the processor could have been able to spot the trend before a large-scale contamination event occurred. Unfortunately in this instance, the control chart was produced from the records in response to an issue several months after the fact. The supplier didn't even recognize an issue was present until the company that purchased the flour notified them of a micrometal problem. Needless to say this had to have been an awkward conversation for the supplier to have with the purchaser. The entire issue could have been prevented with a minimal amount of effort by setting up a daily control chart to be reviewed during the existing record review process.

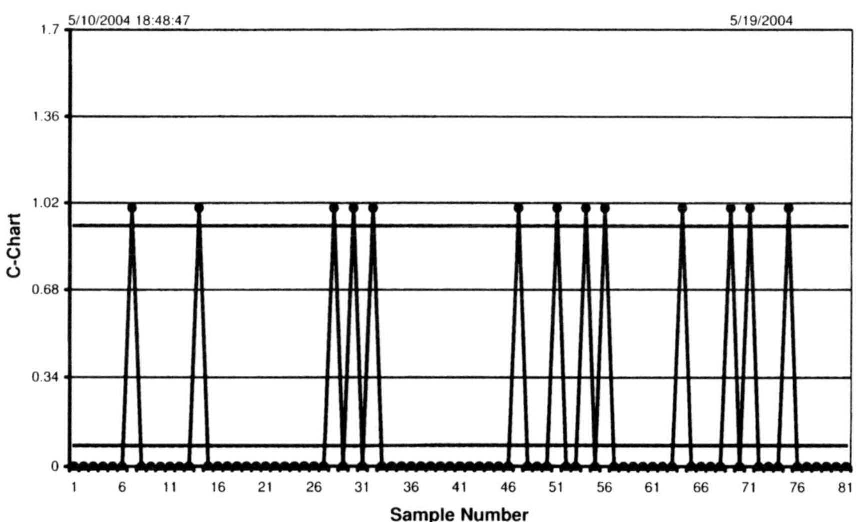

Figure 9.14. C-chart of trace metal found on magnetic separators at a flour supplier to the food industry.

9.9.2 Rejects from automated inspection equipment (legitimate and false)

Correct strategies for specifying inspection equipment purchases were presented in earlier chapters. For this example the points must be presented again for the sake of clarity. When purchasing automated inspection equipment for FM (metal detectors, x-ray, vision systems, etc.), the false reject rates and legitimate limits of detection must be explicitly stated, measured at final installation, and quantified prior to operation. Most processors and vendors do quite well at setting up inspection equipment in the manner outlined above during installation; unfortunately, once the inspection equipment is placed in a status of ongoing operation, due diligence around several of these points tends to erode. The data that is routinely collected by the processor within the operational life of the inspection equipment tends to gravitate toward the verification of rejection of known bad standards. Accepting the ill-founded logic of "the detector is only working properly if, and only if, it kicks out the test samples" can result in a large negative financial variance for an operation. This is especially true when the false rejections occur at the point in the process where

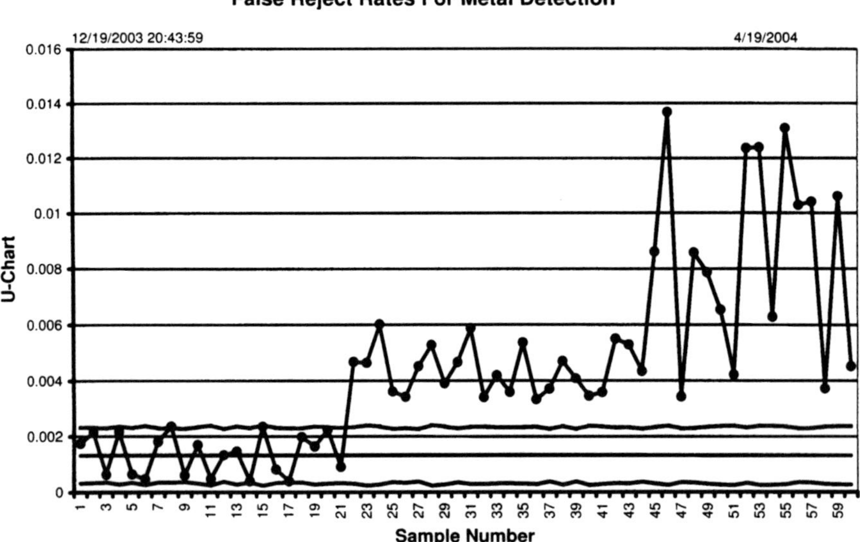

Figure 9.15. U-chart of false reject rates found in a typical application of an automated metal detection unit.

a finished product is present. Using SPC to track the overall reject levels as well as the percentage of false rejects can protect the processor financially. Also, remember that product effects that cause false rejects can also mask true contamination; therefore, due diligence in this area is warranted.

Incorporation of spot attribute trending of false rejects over randomly chosen days of production is usually adequate for detecting a gradual increase in false rejects. If resources are abundant it is recommended that daily monitoring logs be kept to flag the issue sooner. A U chart is presented in Figure 9.15 as an example of daily trending activity.

9.9.3 Results from microscopic analysis of product

It would be laborious and impractical to microscopically evaluate finished product at a number that could characterize a lot with high statistical significance. Practically speaking, the preparation and reading of a sample for microscopic analysis is a process that is time-consuming, and absolute interpretation of the data obtained must be performed with caution. The results of microscopic product analysis are very much dependent on the level of expertise of technicians preparing the samples and using the scope. Due to the variable repeatability and reproducibility of methods involving the microscopic evaluation of products for FM, trending results over time provide greater value, statistically speaking, than evaluating any single or small group of results versus an arbitrary specification. Obviously, finding large amounts of microscopic nonnutritive foreign material in any inspection of a food product is not a good thing, and should always be taken very seriously by the processor. Each process and each product, however, will possess a baseline of microscopic impurities that can be well characterized over time using SPC.

Categorically monitoring these microscopic impurities over time with attribute SPC can generate a wealth of data on the inherent baseline purity of the product or ingredient in question. Given that the technician base generating the analysis results stays relatively constant, large deviations from the accepted "normal" purity level can be flagged using a control chart. Figure 9.16 details a yogurt manufacturer's microscopic monitoring results for its Georgia Peach flavor. The product contains lightly pureed pitted peaches purchased from a supplier as the main fruit ingredient. Evidently an aberration in the supplier's preparation process must have

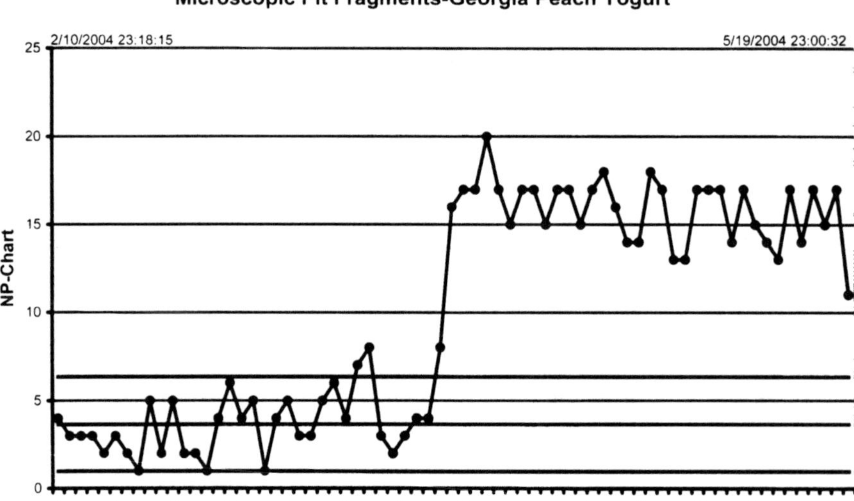

Figure 9.16. NP-chart of pit fragment counts found in the finished product Georgia Peach Yogurt.

recently occurred. Notice the excessive number of pit fragments found in the finished product over the last three runs!

9.9.4 GMP violations in sensitive areas

GMP violations are never a positive event. This is especially true in more sensitive manufacturing operations where products are exposed to the environment, and subject to contamination. It is critical to monitor GMP compliance in such areas regularly, and act with a fast and heavy hand to deal with recurrent issues. One method to help identify an impending system failure is to use SPC to track violations from the audits conducted within the area. If personal practices, such as employees keeping their personal effects in a work area (a common source of FM), have become repeat offenses in area audits, then SPC might help to resolve the situation. Audits could continue to be conducted on a random frequency, but the results can be charted in an NP format. Assuming the individuals implicated had been trained in the relevant policies, a simple visible posting of the results versus the goal in the area is a good first step to driving expectations. Just displaying the importance of the GMP policy to the employees in an area may be enough to make the problem decrease in severity, but if further intervention is necessary baseline data will be present in the form of the NP chart to track the effectiveness of any new management practices (i.e., rewards for improvement, extensive discipline, etc.).

9.9.5 Trending consumer feedback

Unsolicited feedback is one of the most obvious indicators of consumer satisfaction. When handled improperly, however, consumer feedback can put a company in turmoil. Improper handling, coding, or storage of consumer feedback will either generate a lot of wasted effort for an organization, or can mask data causing an organization not to react to potentially serious situations involving FM. Assuming that the mechanism for receiving the feedback is valid (i.e., trained phone rep, unfiltered e-mail, or regular postal correspondence), and the data is processed in a timely manner, SPC is a good way to monitor complaint categories.

Foreign material is always a good complaint topic to categorize because every valid complaint has a root cause that is to some extent actionable by the processor. Figure 9.17 illustrates a consumer complaint trend

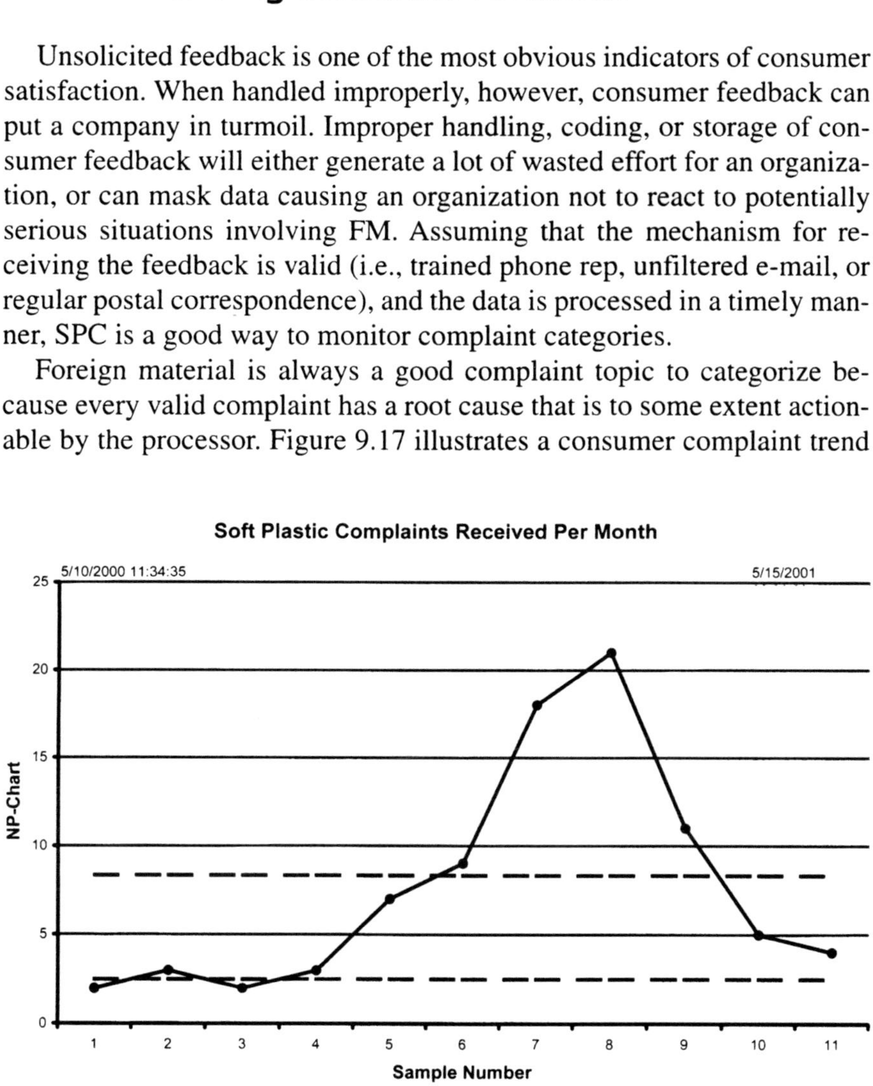

Figure 9.17. NP-chart of an 11-month consumer complaint trend for soft plastic contamination observed in a ready-to-eat meat product.

for a ready-to-eat meat product—precooked and sliced chicken breast. Note the four-month baseline of minimal complaints prior to the unexpected rise in May of the same year. Through monitoring of this control chart, the processor was able to make a quick correction to the root cause of the problem and isolate the suspect inventory before the reputation of the brand was destroyed. Though this is not an example of preventing FM contamination by any means, it does illustrate the broad utility of SPC for supply chain surveillance. In this case, SPC was used by the processor to characterize the product all the way to the consumer's table. Consider it a best practice, and proper due diligence, to utilize SPC for monitoring consumer complaint data.

Works Cited and Select Resources

1. Hubbard MR. 2003. *Statistical Quality Control for the Food Industry. 3rd ed.* New York, N.Y.: Springer. 343 p.

2. Juran JM, Gryna FM. 1970. *Quality Planning and Analysis.* New York, N.Y.: McGraw-Hill, Inc. 684 p.

3. Moore DS, McCabe GP. 2002. *Introduction to the Practice of Statistics. 4th ed.* Cranbury, N.J.: W.H. Freeman and Company. 854 p.

4. *Sampling Procedures and Tables for Inspection by Attributes.* 1989. MIL-STD-105-E. Washington, D.C.: U.S. Government Printing Office. 75 p.

5. *Sampling Procedures and Tables for Inspection by Attributes.* 2003. ANSI/ASQ Z1.4. Milwaukee, Wis.: Amer. Soc. for Quality Control. 94 p.

6. Schilling EG. 1992. *Acceptance Sampling in Quality Control.* New York: Marcel Dekker. 816 p.

7. Steel RGD, Torrie JH. 1980. *Principles and Procedures of Statistics: A Biometrical Approach. 2nd ed.* New York, N.Y.: McGraw-Hill Inc. 672 p.

CHAPTER 10

Management and Communication of Risk in Situations of Crisis

10.1 Introduction

Development of a risk assessment, management, and communication strategy is an essential component to remaining viable as a food processor. As stated earlier in the text, large processors usually do not have inexpensive recalls. Understanding what it takes to successfully manage a risk situation up to and including a recall is a prerequisite for long-term business success. This chapter is dedicated to providing the reader examples of how not to handle situations, and then build a foundation for success by illustrating strategies based on world-class principles.

It is inevitable that a brand will face FM incidences that will quickly escalate if poorly handled. The level of escalation to be expected could include regulatory agencies, attorneys, and the global media. Through focused diligence and adopting thorough organizational policy, situations many times can be assessed and remedied before large-scale organizational trauma is produced.

10.2 Fictional case study: Lucky Flakes, the iron-fortified breakfast cereal

Lucky Flakes was founded 50 years ago by John "Lucky" Mathis, and its Lucky Flakes cereal has been a staple of the American breakfast diet for decades. Through clever marketing and brand positioning, Lucky Flakes had just achieved a 60 percent share of the market, taking back lost sales from private label cereal manufacturers that often position their cereals at a lower price point than Lucky Flakes.

Lucky Flakes has been hailed by the consumer for years as a value-added product due to their total nutritional portfolio and excellent reputation for quality. Recently the growth of Lucky's portfolio has afforded the addition of new staff positions, and Jan Blakely joined the company as the new Director of Quality for Lucky Flakes cereal. Jan started with the company about four months ago, and has been busy learning the ropes for managing product quality at Lucky. Jan's previous experience as a quality director in the nutrition bar business has prepared her well for the exciting new opportunity at Lucky, and she has already made many changes to their existing HACCP and foreign material control programs that have had positive impacts.

Late one Friday afternoon, Jan received a frantic call from the supervisor in the Customer Relations Department. It seems that two people called the complaint line on Friday morning complaining of metal contamination in the Lucky Flakes cereal. The first complaint was received from an elderly man with a heart condition in Elkins, W.Va. While eating Lucky Flakes that morning for his breakfast, as explained by his wife who called in the complaint, he had lacerated the roof of his mouth and injured his esophagus after biting into and accidentally swallowing a metallic object in the cereal. The man was extremely upset, and was so distraught that his wife had to take him into the hospital for fear that his heart condition had become aggravated. The wife had stated that the doctors in the emergency room had removed a 19 mm section of what resembled a curlicue metallic contaminant via endoscopy of the man's esophagus, and a 2 mm sharp shard of metal that was imbedded in the roof of his mouth. Having dealt with these situations before, Jan listened carefully and took good notes. She asked the customer service representative (CSR) if they had the lot codes for the cereal that was implicated as the likely culprit. The code given by the customer service rep was a partial code that only had the manufacturing plant, production date, and an identifier for what line the product was manufactured on.

Jan then asked the CSR about the nature of the second complaint. The CSR explained that the second complaint was received four hours later, and originated from North Bend, Wash. There, a three-year-old was enjoying a nutritious breakfast consisting of Lucky Flakes cereal, milk, and a bottle of apple juice. The mother of the child, who called in the complaint, stated that she could hear her child screaming from the other room where he had been eating the product unattended. As she ran into the room she noted that the child was lying on the floor crying and a trickle of blood was coming from his mouth. Concerned, the mother rushed the child to the emergency room of a nearby hospital, where the doctor had removed a 10 mm × 1.5 mm "needle like" jagged shard of metal from the child's gum line. According to the doctor, the metal appeared to resemble surgical stainless steel. The child received five stitches to close the wound, and a series of injections to prevent secondary infection. When Jan asked the CSR for relevant information relating back to this situation, she was able to produce a full-lot code, and the date of production was the same as the original incident.

The supervisor explained to Jan that she tried to call earlier, but could only get to Jan's voicemail. Panicking, she then called her boss Mark, the manager of customer relations, to let him know there may be a potentially serious issue with the Lucky Flakes. Jan thanked the supervisor for the information, and told her to keep her posted of any new developments or complaints received by the 800 number. Jan then went to her company-provided phone list and proceeded to call the QA Manager of the Lucky plant in Bowling Green, Ky., in an effort to try to understand if there could be an assignable cause for these incidents. The six-digit speed dial Jan dialed did not work, and she was very disappointed to hear the dreaded buzz tone that is indicative of dialing an unsupported extension. Two months before, Lucky had changed phone service providers, and the qualifier extensions (the first two numbers) for the speed dial had changed to include access to their sales offices and international manufacturing facilities. Dismayed by this, Jan thought quickly and dialed Information to get the phone number for the Kentucky plant. She was connected to the plant by the operator, and to her dismay was put in contact with an automated answering service. Precious time on a Friday afternoon was ticking away, so Jan began to press the 0 button repeatedly in the hope of reaching a human. Jimmy, the on-site security officer, answered Jan's call, and transferred her to the QA manager, Mr. Arlie Land. Jan, being relatively new, had rarely spoken with Land up to this point so the Friday afternoon call had caught him completely off guard.

Jan proceeded to explain the news to Land, who was quite distraught over the situation, and said that he would proceed immediately with a records review and line walkthrough. This was a prudent first step, seeing that the product was manufactured less than a month before. Land did allude to the fact that he remembered that the plant was switching over to new state-of-the-art metal detection equipment just prior to the complaint date. He was quite confident that the metal implicated was not a plant problem. Land promised to get back with Jan as soon as possible following the review to let her know if any issues were identified.

Jan, now knowing that at least the fact-finding process had been initiated, opened up the Procedures Manual given to her as part of her systems orientation. To her surprise the first two tabs of the manual were missing most of the pages. Being a seasoned veteran of these situations from her nutritional bar days, she figured that she would have to wing it and do the best that she could under the circumstances. Jan then attempted to call her boss Bob, the Vice President of Quality, to fill him in on the two complaints. Unfortunately Bob did not answer the phone call, and Jan remembered that he was to leave early Friday to embark on a trip to Vancouver, Canada. She left Bob a voicemail explaining the situation and eagerly cleared the line awaiting a call back from Land.

At 6:00 p.m. Land called in frustration and explained to Jan that he had conducted a review of the records and found that the metal detector verification paperwork was not in the pack record for the complaint day. He also explained that the day in question was, in fact, a record production day of more than 250,000 boxes. Jan asked if he could begin a product trace for that day through initial distribution using the company's new multimillion dollar enterprise system. The Integrated Distribution and Operations Network Overview system (IDONO) had been recently implemented at Lucky. Land laughed, and stated that he had not been officially trained in the use of the new system, but he would begin a paper trace while calling someone in Information Technology (IT) to get data from the IDONO system.

The paper trace took until Monday morning to get completed, and by that time Jan had finally been able to bring the management team together for a briefing. Unfortunately several of the team's representatives were not able to be present due to business travel and vacations. The most notable vacancies were those of the corporate attorney and the head of logistics. (Bob, the Vice President of Quality, was able to return from Canada in time.) In this management team briefing, the two cases were discussed as well as the lack of any relevant metal detector challenge records. The management team was in turmoil over what the scope of this problem really was, and Bob stated that he considered any product manufactured on the complaint day suspect. The group phone-conferenced Land for a quick recap of any further developments at the Kentucky plant.

The additional information presented by Land was the fact that there was maintenance performed on the filling line the day prior to the date in question, and that some of the product from that date code (15,000 cases) was still in the local warehouse due to a mistake in shipping rotation. He further explained that the scope of the maintenance activity performed was to replace a screw conveyor that fed the cereal to a bulk receiving bin. This bin fed several filling lines that day, including the line that the Washington complaint box was from.

This information really muddied the waters for the group, and after Land was off the phone line the dynamics of the situation really began to flounder. "How can we not know where this product is in our systems yet, or where the initial distribution was . . . it's been 72 hours!?!" exclaimed the company president. The president then leaned back in his chair, took a deep breath, and began to rant again, "I've authorized millions of dollars to renovate all of these legacy systems to IDONO, and we are still relying on a supervisor in Kentucky to dig through a pile of papers stuffed in filing cabinets! We need data NOW, not when IT can get around to it!" Jan shrank in her seat in discomfort as the Chief Intelligence Officer seated next to her gave out an audible grunt of dissatisfaction. Obviously this was not a good day to be in quality assurance or IT.

Bob, looking at the complaint files, stated, "Until we can get a feel for the initial distribution we are at a standstill for managing the long term. We can, however, take a pre-emptive strike and order all of the inventory from this production date within our immediate control to be placed on hold status." Phil, the Vice President of Sales, abruptly responded, "Bob, you're kidding me, right? Lucky Flakes have been flying off the shelves since our last vitamin improvement, and we are selling at 110 percent of forecast. We're fortunate to have one week of forward coverage in our supply chain right now to cover our existing orders, and over the last three months, we have been horrible at servicing our customers! If we hold that inventory, we will probably short half of the Midwest region. That would really make Megalo-Mart, our largest customer, nervous, and I'm sure they would wonder about our reliability too. All of this for two consumer complaints that we haven't been able to fully link together yet . . . what's the significance, Bob? Our salespeople are out there trying to drive home the quarter, and you people want me to tell them that they won't be able to deliver? If Brian were here {the attorney}, he would tell you that you guys are totally off base here!"

Jan, seeing Bob's forehead starting to resemble a vine-ripened tomato, decided to interject some logic. "Phil, you need to understand that the company could be in a situation of high risk right now. The swifter we can take action, the quicker we can hope to control the scope and severity of the issue. I know that the business has seen some tremendous growth over the last few quarters, but we have to be sure that we can protect the long-term health of the brand as well as the well-being of the consumers. I think we need to get back to following the crisis management procedures, and let's start by looking at the true risk presented. The retrieval service should be delivering the samples from the consumers' homes this afternoon, and this will allow us hopefully to identify if this truly is our problem. If your group could pull together some sales velocity numbers and start to conduct shelf audits to check for this code, we might be able to better determine our exposure quite quickly. In the meantime, I support Bob in holding the inventories on hand so we don't potentially further poison the well."

At that point the company president had had enough of the debate and sternly began to address the group. He thanked Jan for her composure and stated, "Phil, get your folks on the data acquisition phase now! I don't want to hear of a single expense report filed in the Accounting Department for any client dinners tonight because we need your people combing the shelves to see how much of this lot is still hanging around. Bob, per my direction I want you to work with the junior logistics officer to lock down any inventory that might even have sat next to that suspect product . . . I don't want to be the next 'former' company president having to give a briefing to the media about how we didn't see it coming. Jan, I want you on the next plane to Kentucky to figure out what could have happened. Also try to investigate for me why we can't complete a routine product trace in 77 hours. I'll call the plant manager as soon as we adjourn and tell him that you are on your way. He will be fully briefed to dispatch any and all resources you need for conducting your investigation. People, let's go! We don't have much time to get a handle on this situation because we are now in what the lawyers call an elevated 'state of knowledge,' and I do not want to get any smarter."

10.3 Development of proactive risk management structure

Every company should be prepared to handle adverse events similar to those in the Lucky Flakes example. Preparation should go beyond the legal requirements of having a recall plan that complies with the relevant regulations governing the industry. Remember that the goal is not to recall products, but to manage situations in a way to prevent any recalls if at all possible.

The Lucky Flakes case study presented above illustrates a company that was for the most part ill-prepared to handle adverse events. For most companies, a lack of preparation stems from lack of prioritization of developing risk management strategies (1 and 4). With success and time, some companies are lulled into the comfort zone of feeling that an emergency situation is so unlikely to occur that they spend time and resources elsewhere dealing with more immediate operational needs. With the advent of food security threats, the "it won't happen to us" culture must not be allowed to proliferate or endure any longer. FM incidents that were once considered remote possibilities for some industries must now be considered reasonably likely to occur.

For these reasons it is now imperative to develop risk management policies and procedures to be used by the organization in times of crisis. It is not enough just to publish books of policies and procedures that are produced via a third party or copied from another location's master doomsday plan. In order for the policies and procedures to truly deliver meaningful results, they must be highly specific to the work processes present within your organization. Generic policies and procedures that offer no real "how to" instructions or role definitions are nearly worthless in all but the smallest companies.

Usually a cross-functional team of personnel from quality, operations, logistics, customer service, legal, corporate communications (if this group exists), and consumer relations is the best group to develop the high-level policies related to crisis management. Realize that risk management situations involving FM incidents usually take some time to resolve. This is due to the nature of ancillary activities that usually need to run their course. Receiving the samples in the laboratory for analysis, conducting records reviews, and identifying the location and quantity of suspect inventory can and does consume precious time. This is true even in companies with excellent risk management practices and world-class systems. The crisis management team must be prepared for long hard days that will keep emotions running high. The leader of the group must be sure to keep the emotional side in check with relevant data. When emotion is kept in check with data and logic, the decision-making process will remain unclouded.

10.3.1 What defines a situation that requires high-level risk management?

The policies that will serve as guidance to the organization in times of crisis will first need to loosely describe the scope and severity threshold that must be met for an incident to trigger the culmination of structured risk management. Prioritization of such issues is absolutely critical—you would not want to call together Executive Committee members for two "hair-in-the-cereal" complaints received on a 10,000 unit production date.

Likewise, when educating personnel within the corporation about the threshold definitions arrived at via consensus, emphasis must be placed on the fact that these are presented as guidelines, not absolutes. This is an important fact because, frankly, sometimes even the tone of the caller on the other end of the phone line or written word contained within the correspondence could be enough to warrant further investigation into a situation. This is especially true when a caller or letter makes reference to tampering or other malicious actions that a group could be planning. Establishment of threshold situations will act to guide the organization into an incident-management mode where the highest priority is given to FM complaints, media reports, or regulatory inquiries that could represent a risk to the brand.

10.3.2 Define who is to be directly involved

Knowing when to report an incident is only part of the knowledge necessary to move forward with a risk management strategy. The "who" question is one that often gets muddied in a potential crisis. The members of the management team that will be on the crisis management team must be clearly identified in policy. It is usually best to cite the function, such as QA Director, rather than names in a policy definition statement . Not only is this a good policy-making practice, but by defining functions versus people the policy will be easier to maintain in the future. There must be clear roles and responsibilities established for the key functional representatives that will be part of the crisis management team (1). Nothing is more counterproductive than when a team member will try to overstep his or her functional accountability in a crisis management meeting. Case in point: when the Director of Sales tries to debate the legality of a nonconforming situation defined by legal counsel . . . this is clearly a case where the situation must be defined by legal counsel, who retains this core competency.

Another common mistake that will be made if not spelled out in a clear consensus policy statement, is that the roles and responsibilities of a crisis management team member must become the ONLY priority of this person until the crisis situation is resolved. Everyone who has access to the crisis management policies must be aware that those individuals whose functions are identified with managing the business risks will be OFF LIMITS for other business matters until informed otherwise.

10.3.3 Defining the other "who"—continuity of business operations

For reasons mentioned above, the other part of the "who" equation is to designate backups for the personnel that will be responsible for chairing the crisis management team. This is important because having functional backups provides a framework so that business-critical functions can continue at some pace, albeit probably not as crisply as would happen in non-crisis management mode. Typically, this exercise would need to go two functions deep. By mapping the policy to this depth, it can be assured that situations such as vacations, sick leave, or employee turnover will not hamper the risk management team's efforts. An added benefit received in adopting this approach is that, from an external viewpoint, only faces and names have changed for a period of time. This can be keenly important in maintaining corporate image because outsiders won't get tipped off that the company is potentially implicated in a situation of heightened risk.

10.3.4 Develop defined communication channels for internal stakeholders

Most corporations handle external communications of crisis situations in a satisfactory manner. There are public relations firms that cater to such situations and can help develop turnkey strategy for the company. Internal communications are, however, many times poorly handled by corporations when working in a mode of crisis management. So many times in situations of crisis, higher-level personnel throughout the organization will start feeling anxious and begin calling the people who are conducting the crisis investigation (4). This is usually done in an attempt to ascertain up-to-the-minute facts about the crisis. It is easy to see how a concerned member of the management team would begin to go down this path—these individuals have a great deal of ownership for the business, and in a vacuum of information they can begin to develop extreme anxiety.

Unfortunately when a barrage of internal information requests hit members of the crisis management team, it usually interferes with their duties. By spending time replying to voicemails, e-mails, or water cooler interrogations, these key individuals are neglecting their duties in resolving the crisis situation. Spending time communicating can delay essential initial fact-finding or containment activities. This can be deadly for an organization when speed and action are an essential component to potentially protecting public health and the identity of the brand. Such situations can be avoided altogether by generating a policy statement that details the structured communication strategy of the organization during periods of crisis management. A supplemental procedure will also need to be developed that outlines the information flow for internal communications. The procedure should remain mute in specifying the frequency of such communication because of the ever-differing levels of communication required when dealing with situations of crisis.

Early on in the management of the crisis, the briefing schedule must be made known to all stakeholders. This will be a very beneficial step that should be considered mandatory in dealing with any crisis situation. The organization will be likely to act in a concordant manner in dealing with the crisis at hand if the crisis management team propagates a single uniform official message at an interval that deters formation of a vacuum. The common discordant approach that is often seen when organizations attempt to crisply react to crisis situations is usually created through fragmented "grapevine" information being dispatched at random frequencies. This is easy to understand because information is constantly being received and revised by those at the heart of the crisis, and personnel on the outside who are trying to be helpful may put forth efforts in areas that are unimportant for crisis resolution.

The content of internal communications must be succinct and written in an "as we understand the situation to date" manner. No conjecture or preliminary data should be cited until it is irreversibly confirmed. The communication should be dated and the crisis team members currently involved should be referenced as well. The preventative approach being employed for the protection of the consumer should be the primary theme of the communication (and should be the primary reason for any crisis management strategy), and a best estimate of the date when the next communication will be sent forth should be identified. It is extremely important to realize that every situation will potentially be re-analyzed by regulators and/or legal professionals after it is resolved, and anything that is in writing can be used as incriminating evidence (e-mails, memos, and documentation of meetings). It is also important to remember that any official communication that is released could be leaked to the public. Simply placing a "strictly confidential" stamp on a document does not offer tangible protection in the court of public opinion where retribution and sensationalism are a fact of life. It would probably be good practice to develop drafts of communication statements that could be used during common crisis situations. Drafts for product tampering, FM complaints received by consumer affairs groups, citation by a regulatory agency (foreign and domestic), implication of wrongdoing by a non-governmental organization (such as Greenpeace or PETA [People for the Ethical Treatment of Animals]), or reaction of the media to an FM incident are typical situations of risk faced by companies.

10.3.5 Ensuring that inventory can be traced, contained, and accounted for

Increasing pressure from nongovernmental organizations (NGOs) and governmental agencies is making traceability from farm to table a necessity for all food processors. Simply having policies and procedures that reference the identification of suspect products and containment of said products is insufficient for protecting your brand or public health. The procedures and systems necessary for identifying suspect products and tracing them back to their origin are fundamental quality system components for any processor. It is essential to identify lots manufactured and the subcomponents that comprise such lots when attempting to contain a potentially significant FM incident. The ability to stop shipment of a product and contain it from being inadvertently distributed is also an important execution point for the processor. When traceability and containment of products suspected of being contaminated by FM is assured, the processor will have at least one level of risk addressed in the bigger picture.

10.3.5.1 Issues faced by processors when establishing traceability programs

An absolutely critical component that is missed by many processors is a fundamental understanding of how their suppliers identify unique lots of raw materials. When dealing with a corporation that is the primary supplier of the material, a simple phone call can usually clear up this mystery. When dealing with brokers, however, many times the lot integrity of the material was lost two acquisitions ago. This is a situation where the purchaser must balance the risk accordingly. Obviously, the decision to institute traceability for any component is a delicate balance of risk and reward, but analyzing the risks in a proactive manner is a far simpler strategy than trying to sort things out retrospectively.

Even if the material producer's traceability is understood, it is important that the processor maintains the chain of traceability throughout the balance of the supply chain. The procedures used to uniquely identify lots of raw materials at the receiving stage of the process is usually the first point of weakness in the traceability program. This is especially true when bulk loads of materials are received. Wet and dry products stored in large bins, tanks, or silos represent a worse-case scenario for maintaining traceability. In such a situation there will be multiple lots commingled in a tank, and each lot received thereafter will be conveyed in on top of them. Careful monitoring and documentation of the receiving dates of the materials, daily usage quantities, and any periods where the bulk storage was emptied or cleaned can sometimes be the only evidence that a lot was ever completely exhausted from such a system. Other common receiving area mistakes that must be accounted for via defined procedures and auditing are:

1. Combining short loads or multiple arrivals into one received lot
2. Not documenting incoming load (or door) rejections or supplier shortages
3. Failing to keep records that are of adequate detail to identify what was received at the facility, and when
4. Poor recordkeeping practices (storage, pencil use, no dates, etc.)
5. Accepting loads without review of relevant certificates of analysis or performing specified sampling

The next common stage of the process where issues occur is usually found in the usage and storage of partial lots. Manufacturers will sometimes purchase dry ingredients, such as a vitamin mix or spice, and use only a very small quantity of the material in each batch. Many times the ingredients will come in a multilayer paper bag with a plastic liner. The outer layer of this bag will contain the identity of the product and its relevant lot number, but it is common practice to rip this outer layer off prior to using it to ensure that the product does not become contaminated by any dust that may have settled on the bag during warehousing. If only half of the bag is used and the information is not permanently attached to the remaining half of the bag, it usually will become lost in the process. Likewise, if the bag is ripped across the code-dated portion and then folded shut for later use, it is very difficult to identify what material lot it originated from.

Rework streams and off-line nondestructive testing scenarios are also notorious for undermining traceability efforts of producers. Anytime intermediate products are taken away from a production lot and reintroduced at a later time, extreme diligence is warranted in the creation of detailed product handling and tracking procedures. These practices must be audited at a high frequency, and put under extreme scrutiny. Dealing with an FM incident is an ugly scenario in itself, but not knowing whether the scope of the incident extends beyond a single day of production can be a sad and sobering reality. If rework and reconditioning is unpreventable in the processor's industry, then assigning a shelf life to the rework intermediate product (one day, one week, one month, etc.) can help to reduce the potential scope of a product issue to a slightly more manageable quantity.

10.3.5.2 You know what it is and where it is, but can you stop it?

Being able to quickly and effectively "lock down" suspect inventory within the immediate control of a processor can prevent a broader market action or stock recovery from being necessary. Conversely, assuming that shipments have stopped and an issue is contained when in reality the product is still being received by a customer, is a potentially sickening scenario. Clear definitions and guidelines of what "on-hold" means must be stressed from boardroom to loading dock. If the truck, railcar, or shipping container that the product was loaded on is still on the property, the processor has the right and authority to take isolation action against the inventory. Obviously certain tact must be used when discussing the matter with the transportation company, but if a potentially brand-threatening situation can be managed, empower those in the area to manage it authoritatively.

Physically placard the material in a manner that readily identifies its hold status, who authorized the product hold, and the date it was placed on hold. This product should be placed in an area that would prevent it from being mixed with good product and shipped. Physical isolation scenarios provide the best results over time, but if an electronic inventory system is used to mark the product as "on hold" within the system, it should be treated as a validated system and audited to ensure ongoing effectiveness. Most of the enterprise systems commercially available for regulated industries have provisions for dealing with product on-hold status, but validating the system "as installed" is warranted.

10.3.6 Prepare, drill, and follow up

Even if all procedures and auditing programs have been created and appear to be working, it is prudent to test the traceability and hold systems on a routine basis. Routinely running "mock events" can keep the organization sharp and identify system inadequacies. If the processor has never performed a mock event, the first one will usually be an enlightening experience. In the infancy of a mock event program, focus on an issue that deals specifically with a product produced on a certain date. Usually this can be done by looking at a facility's production records, product-retain library, or a store shelf for inspiration. Try to choose products that would have been through initial distribution to ensure that the process is sufficiently challenged. When running the drill, ensure that the organization treats it as if it were the real event and continuously emphasize the importance of taking it seriously. Set a stern time limit for the event to be completed, and then prepare to observe the system in operation. The hold system will need to be challenged as well. This is a controversial step, but it is the only way to ensure that the system will work in times of crisis. Shorting customer orders is never popular; as long as the event can be resolved in a day this situation can be avoided. Overall, the organization should be able to track 100 percent of the product implicated through initial distribution . . . anything less is unsatisfactory.

The time it takes to perform a mock event and the amount of product traced and held are the key performance indicators of the systems. These measures need to be trended and reported. As the organization begins to improve on the basic mock event strategy, a more advanced drill can be introduced. Once again, select a product and manufacturing date to be the subject of the "mock event," but this time stipulate that one of the lots of ingredients used to manufacture it has been implicated in an FM incident. Performing this type of drill can really flush out program inadequacies from poor purchasing strategies to mishandling of raw materials. Bulk spice lots and processing aids will usually represent a situation of usage over a very large amount of finished product. In one case, a single lot of spice purchased by a major corporation was enough to last for 18 months of production . . . that's truly putting all your eggs in one basket.

As with any other system, it is important that your system be able to sustain its function even when key individuals change positions or move on. Ensuring that all critical functions are drilled two levels deep will ensure that there is at least exposure to the nature of a "mock event." The sports analogy of "how you practice is how you play" is very appropriate in crisis management situations, so a certain degree of bench strength will be required. There are very few things that can be worse than hearing the person on the other end of the phone line sheepishly state, "I've never done this before, but I guess I can try to get what you are requesting" when information is necessary to resolve a potential crisis situation. Education of key personnel, drilling of the systems, and ensuring that the system works will prevent this situation from occurring at times when calm and timely execution of tasks are absolutely essential.

10.3.7 Defining authority levels for product disposition decisions

It is extremely important when dealing with crisis management situations that there is an end to the process. In many situations the end of the process either results in some action taken with the implicated product (recall, market withdrawal, stock recovery, etc.) or no action being taken with the product. There needs to be clear policy on who can authorize these actions. Committees work wonderfully for gathering facts and coming to consensus with recommendations, but there needs to be an individual who has ultimate authority to identify when a recall is or is not justified. This is extremely important, because in very large global organizations regional business heads may, unknown by senior management, execute local recalls and product withdrawls. The ever-present global media can make this practice quite deadly for an organization's established brand equity. For this reason a single point of authority must be defined within the organization, and the accountability will also be centralized for product actions.

Usually the person best suited for handling this task is the business owner, operating president, or Chief Executive Officer (CEO) of the corporation. Obviously, for execution of stock recoveries and other low-level practices that may affect nondistributed inventory, the approval could originate from lower levels of the organization, but when making the decision to initiate a publicly announced recall there has to be approval and knowledge at the highest levels of the organization. Many would argue that, from a legal culpability standpoint, it would be better to leave the decision in a consensus form among functional heads, but there are cases where the head of a corporation can still be held liable in criminal or civil court for decisions that are made by the organization without his or her knowledge. Not only that, but fighting over aspects of plausible deniability does not help protect the consumer or investor from the effects of an FM issue!

10.4 Pulling the trigger and recalling product to protect public health or the reputation of the brand

Obviously there will be situations that will justify a recall be initiated to protect the health of the public or save the brand from certain ruin. When the time comes to perform a recall, the state of readiness of the systems proactively created will be given their ultimate test. Drilling up front and ensuring that folks are trained on the latest policies and procedures will provide crisper executions, but many of the results that define what a successful recall is are situation-dependent. This section of the chapter will present the recognized definitions of what a recall is, and some of the commonly forgotten subtle points when executing a recall. The last section of the chapter deals with communicating market actions to the public and profiles several exemplary best practices adapted from companies that have successfully survived times of crisis.

10.4.1 Understanding relevant federal regulatory agency recall definitions

Before executing a recall or market withdrawal, it is important that the processor understand the requirements of the regulatory agency that has jurisdiction over the product involved. Both the FDA and USDA maintain recall classifications that spell out the scope and requirements necessary for conducting a recall. Realize that the strategy that is allowed to be taken to resolve the recall situation differs based on the classification of the recall events. It is also important to realize that, even if you, the processor, elects not to recall a product for a specific situation, the regulatory agency may find your reasoning unfounded and take action against the product in commerce.

Additionally, if a recall matches one of the defined classifications and is in fact performed, the agency may decide that the level of recall executed by a processor was not adequate. This could result in a demand or a request that the processor apply the more stringent tactics required in the next levels of recall execution. If that happens, publication of these recalls would be changed in the respective recall databases maintained by both agencies. Listed in Table 10.1 are the recall classifications and definitions

Table 10.1 Regulatory agency definitions for market actions taken upon food products distributed to the trade and consumer level.

Type of Market Action	*FDA Definition**	*USDA Definition***
Class I	A situation in which there is a reasonable probability that the use of, or exposure to, a violative product will cause serious adverse health consequences or death.	This is a health hazard situation where the use of the product will cause health consequences or death.
Class II	A situation in which use of, or exposure to, a violative product may cause temporary or medically reversible adverse health consequences or where the probability of serious adverse health exists consequences is remote.	This is a health hazard situation where there is a remote probability of adverse health consequences from the use of the product.
Class III	A situation in which use of, or exposure to, a violative product is not likely to cause adverse health consequences.	This is a situation where the use of the product will not cause adverse health consequences.
Market Withdrawal	The product involves a minor violation or no violations.	A firm's removal or correction on its own volition of a distributed product that involves a minor infraction that would not warrant legal action by FSIS and constitutes no health hazard.

* Definitions for Class I to Class III recalls taken from FDA IOM Chapter 8. Market withdrawal definition taken from Chapter 7, Recall Procedures, of *Regulatory Procedures Manual*, March 2004.
** Definitions taken from USDA FSIS Directive 8080.1, Revision 4, May 24, 2004.

cited by the USDA and FDA. Note that for each classification the recall can be as narrow in scope as internally held stock, or be as far reaching as to the level of the consumer's home.

10.4.2 Market withdrawals—a separate issue

Both regulatory agencies recognize that, if the producer removes product from commerce and the product in question did not represent a public health risk or legal violation, this is an allowable practice known as a market withdrawal. The USDA definition of a market withdrawal cited in FSIS Directive 8080.1 at the time this text was assembled is:

> "A firm's removal or correction on its own volition of a distributed product that involves a minor infraction that would not warrant legal action by FSIS and constitutes no health hazard."

The FDA interpretation of a market withdrawal is slightly different, and is stated for internal purposes in their *Investigations Operations Manual* as:

> "A market withdrawal means a firm's removal or correction of a distributed product which involves a minor violation that would not be subject to legal action by the FDA or which involves no violation; e.g., normal stock rotation practices, routine equipment adjustments and repairs, etc."

There are also provisions stated within this same definition for tampering incidents:

> "When an investigation determines there is no evidence of manufacturing or distribution problems, but a firm has removed products from the market as a result of actual or alleged tampering with individual units, the action will be considered a Market Withdrawal."

10.4.3 Agency authority to request a recall or seize product

It is important to note that there are differences in practice afforded to the regulatory agencies related to requesting recalls and seizure of products. The FDA and USDA can request that a food product be recalled by a processor, but unless the product in question is classified as an interstate milk shipment or infant formula, the FDA cannot mandate the recall of a food product. The USDA FSIS, however, can legally detain any product found in commerce that its Health Hazard Evaluation Board (HHEB) would deem representative of a significant risk worthy of a recall. In this case, the FSIS Congressional and Public Affairs (CPA) office would issue a press release informing the public that a product that appears to be adulterated or misbranded has been shipped by the responsible firm, and that the firm has refused to recall it. Obviously this is not the sort of public relations strategy that a reputable company would want to use.

The FDA is not powerless from this standpoint either. The FDA can issue seizures against products. This is a slightly more involved process than USDA's, but nonetheless just as effective. The seizure powers of FDA are far reaching and can even include meat, poultry, and egg products. The following excerpt from the FDA's Regulatory Procedures Manual (RPM) illustrates their authority,

> "The Federal Meat Inspection Act (MIA) as amended by Public Law (P.L.) 90-201 and Sections 19 and 20(b) of the Poultry Products Inspection Act (PPIA) as amended by P.L. 90-492 and Sections 19 and 23(d) of the Egg Products Inspection Act (EPIA) provide certain detention powers.
>
> The detention authority under the MIA and the PPIA provide that FDA representatives may detain articles subject to these acts if they are outside a U.S.D.A. inspected plant and there is reason to believe that the products are adulterated or misbranded under the Federal Food, Drug, and Cosmetic Act (the Act).
>
> The detention authority under the EPIA provides that FDA representatives may detain products subject to that act if the products are found outside a U.S.D.A. inspected plant and there is reason to believe that the products are in violation of the EPIA."

The recent advent of the passage of the Public Health Security and Bioterrorism Preparedness and Response Act of 2002 (the Bioterrorism Act) has further expanded FDA's seizure authority. Under Subtitle A of the new act, the FDA was granted authority to administratively detain foods and food products. The scope of this detention authority is still under contention, but rest assured that the FDA will be able to deal even more swiftly with suspect products than it has in previous decades.

10.4.4 Agency authority to reclassify recall status and public notification

The processor must realize that any situation where product is to be removed from commerce beyond their immediate control due to an FM issue could be interpreted as a Class I, Class II, or Class III recall by a regulatory agency. Extreme tact must be used when a market withdrawal or a recall is conducted and the processor informs the agency about it after the fact. If all of the information and facts are not in order, there will surely be a lot of questions to answer, and you can be sure to expect on-site visitation from the regulatory body regardless. This could come in the form of an In-Depth Verification Team (IDV) sent by USDA FSIS, or a team of inspectors from the FDA's field offices showing up at the manufacturing location. Be prepared to exhibit all relevant data, samples, pictures, a timeline of events, and be able to explain the root cause of the problem as well as its resolution. The processor must also be ready to discuss and present any documentation of actions taken to dispose of or recondition the recalled products. If the product was destroyed, there must be documentation of the quantities landfilled; a good practice is to have photographic evidence of the destruction. This can be done as a supplement for USDA products where the Inspector in Charge (IIC) will most likely accompany the product to the landfill and witness its destruction.

Both the USDA and FDA use teams of experts to review the relevant data and facts that have been gathered from inspections, complaints, reported illness, or provided by the processor. The FDA will use a group of personnel (usually from CFSAN) to perform a Health Hazard Evaluation (HHE) on the product in question. This process is standardized, and can be found in Chapter 7 of their RPM (freely available for download via the FDA website). External experts may be brought into the HHE to aid or clarify certain obscure facts, but in the end the situation is given a recall rating based on the HHE. The FDA will post the recall on their website for public review. Generally the FDA does not issue a press release for most recalls as long as the producer is able to issue an adequate press release of their own.

The USDA FSIS uses a similar model for assessing the severity of a situation. The USDA Recall Committee will use an ad hoc committee (HHEB) to determine the risk to public health that the product poses per Directive 8091.1. The potential areas reviewed by the HHEB are explained in Attachment 2 of FSIS Directive 8080.1. According to the FSIS directive, the minimum membership required for an HHEB will include one physician, one epidemiologist, and one representative from the Office of Policy, Program Development, and Evaluation. This group will weigh the evidence at hand and provide a classification for the recall. Based on this feedback, FSIS will post a recall notification report on its web page. If the recall is classified as a Class I or Class II action, FSIS will issue a press release to the general public.

10.4.5 Develop and communicate a vision of what success is

The company performing a recall should possess a shared vision of what will constitute a successful result. As with any other business endeavor, throwing various and costly resources at a project that has no scope can be a dangerous situation. The vision referenced here is far deeper than the criteria necessary for complying with the needs of the regulatory body for completing the recall. That type of achievement is only centered on compliance and, to some degree, preservation of public health. Obviously these are required prerequisites to recalling product successfully, but the vision should be farther-reaching in terms of protecting the long-term survival of the company.

Once a vision of success has been established, it is time to develop a set of shared goals to ensure that the result is delivered. The goals that need to be set to meet the vision should be measurable outcomes that are achievable. Outlining a financial containment objective, such as "execute a recall for under $2 million," is an appropriate goal. Likewise, goals centered on maintaining customer service, speed of recall execution, and market share preservation are all good areas on which to concentrate. Publishing these goals and holding all involved jointly accountable for achieving them provides a tactical short-term focus for folks other than simply succumbing to panic or despair.

10.4.6 Internal communication strategies

In any time of crisis, information will fly via the grapevine at an alarmingly high rate. The accuracy of most of this information is usually only partly true, and the rest is purely conjecture. When active coverage by the media is involved, this quagmire of information gets far murkier. Nothing can be more paralyzing for an organization in the midst of a recall than when employees who have roles in either executing the recall or preserving what's left of the existing business lose focus due to an information vacuum.

Great companies are composed of great people, but even great people tend to get quite anxious when their corporation is in the throws of a recall situation. It is common knowledge that there have been many large food companies that have suffered recalls of great enough significance to result in plant closings and personnel layoffs. People at all levels need to be kept informed of the latest developments and given the relevant facts necessary to maintain their morale. These employees are going to be the greatest assets the company has for the rebuilding phase, and those who are not engaged or are lacking in morale will not pull their weight when times are tough. Corporation employees should not be relying on the media for the latest information involving their livelihood. This is a very poor practice for a management team to employ, and can yield greater issues if the media were to start asking employees for statements as they are leaving the facility or wandering through the community. To the layman, any person who works in a food plant is an "expert" in the safety aspects of the products made by their company. Just imagine how reassured the public becomes when they see a person standing in uniform being interviewed about a widespread recall, and all the employee can say is, "We really know nothing about that . . . they don't tell us stuff . . . but I'm not surprised they aren't saying more to you."

Internal communications should not differ in the message from the external communications, but any extra relevant detail should be shared with internal personnel. These statements should be issued well ahead of any official press releases, and be marked as confidential. The language should be specifically written (7th grade level) for the employee audience. When making such a broad communication, realize that all company personnel need reassurance that every possible scenario is being examined to protect the consumer while work in parallel is being performed to restore the brand image of the company (8). With their help and continued work, the company will soon be well on its way to a complete recovery.

10.4.7 External communications

External groups that are commonly addressed via external communications in a public recall situation are:

1. News media
2. Shareholders
3. Trade organizations
4. Customers
5. Consumers
6. Government agencies

When developing external communications, it is extremely important to use tact and guidance learned from other organizations that have survived recalls. There are several key concepts that the processor must take away and internalize.

The first concept that must be learned for adopting external communication strategies is the nature of public opinion. The general public in most cases presumes the processor to be the guilty party in a recall situation, and presumption of innocence must be earned. The public, for the most part, does not care whether the actions that produced the recall situation were legitimized by the law, trade agency guidelines, internal specifications, or industry best practices. The public skims over such data and facts presented in communications, and filters the situation for morality. If they perceive the company's communications to depict greed, immoral conduct, speculation, internal incompetence, callousness, or an attempt to deflect responsibility, they will become quickly disenchanted with the company.

Successful communications are launched very quickly after an incident has been uncovered, and engage the consumer at a level that displays the company as personable and empathetic. Written communications should come personally from the company president or CEO, and any live or taped interviews should be conducted by these same individuals or other high-level officials who have had media training. Scientists and other company "experts" usually are poor choices for spokespeople unless they have been extensively prepared and given media training. Remember, many of the people that are going to be viewing or listening to the interview may have never even seen a farm or agricultural operation, let alone the inner workings of a food processing company. The public can easily relate to the fact that, whatever the foreign material was that was in the product, it should not have been there in the first place. Biological and chemical contaminants are generally poorly understood by the public, but one would be hard-pressed to find a consumer who does not know what a shard of glass or a metal needle looks like. When recalls happen for contaminants of this nature, the public will rightfully want to know, "How in the world did they let that get into the product?"

The communication presented should not contain technical jargon in an attempt to try to win over the public. Again, it is important to remember that the members of the general public are not experts in hazard analyses or food manufacturing. A typical statement to avoid is, "The products in question met all specifications and regulatory requirements at the time of shipment." (7) Even if the statement is true, the public would most likely hear a message of denial, or believe that the company was hiding something. The communication should also not make reference to others, such as suppliers or distributors, potentially being at fault (think of the Ford Firestone tire debacle here), unless it is a known external tampering event. Other communication guidelines and tactics to follow are listed below:

1. Disclose all of the facts known at the time, but no speculation to the obscure (7).
2. Express utmost concern for the safety and well-being of the consumer, and urge the public to immediately discontinue use of the product in question and to communicate to others to do so as well (3, 1, and 4).
3. State multiple times what products are affected, and be very specific about detailing the unique codes and the method that the consumer can use to receive immediate dispensation for returning the "bad" product (4).
4. Reassure the consumer of the definite safety of any other products manufactured and distributed by the company that aren't implicated in the recall.
5. State aggressively that the company is determined to find the exact cause and apply preventative action at all costs, and there is no timeline that is more important.
6. If only limited facts are known, inform the public that an outside investigator will be engaged to help clarify what could have caused the issue to occur.
7. Note the previous good standing the company has had in times of business and express a sincere desire to regain their trust.
8. Project a charitable and caring image for those who may have been injured; offer to donate resources to prevent the root cause from affecting other food manufacturers; and offer to testify to regulatory agencies and/or government officials about the experience and the root cause (3).

10.4.8 Actively engage with and provide timely feedback to the regulatory agency

The regulatory agencies spell out the minimum strategies for compliance, but going above and beyond these in clear view of the regulatory agency will aid the processor. Maintaining a cooperative yet challenging position with the regulatory agency can result in ensuring that the processor's point of view and relevant data are correctly interpreted. This will require a dedicated resource and perhaps create a hole that needs to be filled by a person of potentially lesser capability on the recall team. The benefit to this approach, however, is that it may decrease the level of recall necessary, and prevent the agency from misinterpreting inspection results or other data that could cause panic and a potential increase in the scope of the agency. Obviously all nonrequested documentation shared with the agency will need to be carefully scrutinized so it is in a format that lends itself to interpretability. Some argue that nonrequested information should never be shared with regulators, but if it helps to aid the processor's cause of narrowing the scope or severity of a recall situation, the action should be considered.

10.4.9 Execute the business continuity plan

A universal tenet of the consumer goods business is that, when the top line is lost, it is not easy to achieve a successful bottom line. Lost sales are difficult to recoup in any period, but it is exceptionally tough following a publicly announced recall. As was discussed in the previous crisis management section, a business continuity plan must exist for managing through this tough period. While the recall team is consumed with ensuring that the product is swiftly brought back from distribution and accounted for, the rest of the business must be focused in other areas. The first order of business is to appoint and empower the "shadow management" team to continue the daily operation of the business. This group needs to be fully empowered to perform whatever duties are necessary to ensure the business survives. This will mean establishing temporary, but official, reporting relationships of personnel. Adjustments to fiscal and fiduciary approval authorities should be quickly established, and all normal business communications should be rerouted to this group.

The first mission of the shadow management team is to ensure that the root cause of Recall A is not going to rear its ugly head in the near future and affect other products that the company offers. Once this reassurance has been reached, the next order of business should be trying to reassure current customers and vendors that the company is still stable during this difficult period. Recalls coupled with mass customer service outages are a sure way to enter a death spiral for a company, especially if a competitor has enough capacity to launch a full assault on your accounts.

For this reason, maintaining outstanding customer service levels are essential. All items not implicated in the recall that can be distributed freely should be distributed, and will be the key to rebalancing the company's finances after the short-term upside-down period elapses. A fire sale tactic of existing inventory is not necessarily the best strategy at this point, because it displays desperation to the trade and can gouge into profits that the corporation will be in need of over the short term. Offering loyalty incentives and premiums, however, are best practices.

The treatment of the consumer is extremely important during this period as well. The shadow management team must ensure that the consumer will be able to receive the same amount of contact/information (if not more) that they were accustomed to prior to the recall. This may mean temporarily expanding call-center and letter-writing capacity and availability. On the electronic side, e-mail attendants and dedicated administrators for the company's web page will need to be increased as well. The consumer must see the company as a strong organization that is not breathing its last breath before their eyes.

Financial and sales reporting during a recall is critical, and the frequency of reporting needs to be increased even if it is at the expense of accuracy and precision. For the shadow management to be effective, they are going to have to know "the score" as events unfold—receiving a report at the end of the month will not allow for the team to employ course corrections along the way. As stated previously, an approximate number will be sufficient because, if the team has to wait for every pallet, penny, or labor hour to be accounted for, the game will be over long before they figure out that it was lost.

10.4.10 Financial forecasting of recall costs

Recall costs can typically escalate when the media is involved. Defined "buckets" need to be set up to account for miscellaneous expenses that will be incurred by the corporation so that proper forecasting can be maintained. Typical costs that are not anticipated and will incur over time include:

1. Reclamation of other nonaffected products
2. Transportation and destruction of returned goods
3. Storage of returned goods
4. Nuisance fees and restocking fees from the trade
5. Insurance carrier fees
6. Legal fees from lawsuits originating from consumer injuries or breach of supply agreements with customers

7. Demands from vendors to receive payment in different terms due to future financial uncertainty
8. Sorting of returned goods due to "good" product being mixed in with "bad" product by stores or distribution centers
9. Obsolescence of current "good" inventory of the same type as the recalled product due to a decrease in sales velocity for the item(s)
10. Unbudgeted use of advertising agencies, communications consultants, and spot writers to launch a recovery campaign or thwart inaccurate or slanted media reports.

By properly accounting for each of these expenses, it will be easier for the business leaders to determine if many of the "big hitters" will be recurring items or one-time events. This is extremely important to understand because the initial recall hit will most likely be a very large financial burden for the company that should dissipate over time. Maintaining a proper forecast of the costs involved will afford the business leaders and shadow managers to start implementing cost containment initiatives to manage long-term and short-term issues. Knowing which costs are one-time only in nature can also make it easier to communicate financially disappointing periods to shareholders. This is because the costs implicated for the recall will not be incurred again (hopefully) in the next reporting period.

10.4.11 Dealing with returned product

In some situations it may be advantageous to enlist the help of an experienced third party to handle and account for product returns. Realize that the regulators will want very detailed accounting of records of receipt, rework practices, and destruction of the suspect product. The general requirements of each regulatory agency when dealing with returned products are detailed within their aforementioned recall directives or policy manuals. Beyond compliance with this, however, the processor is going to have to elicit certain elevated measures to ensure the business remains viable.

Some underhanded stores and distribution centers will use a recall period as an excuse to empty inventories that were nearly obsolete or overstocked. Be wary of this practice, and take measures to protect the company from such fraudulent practices. Using on-site auditors at the point of receiving is a great way to inventory the incoming materials. To ensure that other shelf-keeping units (SKUs) are not sent back, a scanning mechanism can be rigged up at the receiving location. Such a system will also help to provide inventory accounting, and rework rates when used properly. To ensure that the correct date codes are being received, the use of statistical-based lot inspections (described in Chapter 9) is another good method to use. Any rejected lots should be sorted with extreme scrutiny, and those customers who are attempting to unload products on the business should be contacted and given specific data to illustrate that the business is keeping a close eye on the returns. Vendor-managed inventory situations will be the least hassle because of the internal sales representative who can help ensure adequate scrutiny.

Extreme care must be taken with returned product to ensure that it doesn't become intermingled with known "good" product. Usually it is best to perform receiving, storage, and rework activities for recalled products in another facility, devoid of that product, to provide the ultimate in physical separation. If the product was recalled due to a foreign material issue that is detectable by a highly specific automated means and can undergo reinspection, then by all means reinspect it. Realize, however, that the code date on the package has been implicated in a recall which, if redistributed as is, could confuse the consumer or folks in the distribution network and generate a second trip back to the facility (see a vicious cycle forming here). The regulatory agency should be approached about recoding the inspected product in a manner that does not break traceability.

If the recall strategy used has been successful, there could still be a high demand for the recalled product. As stated previously, customer service is extremely important during this period, so scheduling of extended production, enlisting the help of a contract manufacturer, and rework of any lots that were held for minor workmanship issues are all good strategies for filling the pipeline. Do not take undue risks or succumb to shortcuts during this period, however, or another recall situation could ensue. The first experience the consumer will have with the product following a recall is a new first impression. If the impression is poor, so are the odds that the consumer will be the source of a repeat purchase.

10.4.12 Conduct a "lessons-learned" meeting

When the dust settles and the corporation has hopefully survived the initial trauma, the recall group should be brought back together by senior management. The purpose of this meeting should be based on continuous improvement, and not be conducted in a confrontational "witch hunt" manner; instead, the processor should focus primarily on what went well. Once the strengths have been extracted, the group can start to focus on more negative aspects of the experience. This discussion must be rooted in processes or systems, and not in the performance of specific personnel. Ideally this will be done with all of the current policies, procedures, and contact lists available for review and mark up. By following up on any inadequacies and systematizing previously unknown strengths that were discovered, the processor will have developed much stronger crisis management and recall execution protocols.

Works Cited and Select Resources

1. Bartlett R. 1999. *Dancing with the Devil: Crisis Management in the Food and Drinks Industry.* Surrey, UK: Leatherhead Food RA Publishing. 262 p.

2. Bierck R. 2002. *What Will You Say When Disaster Strikes?* Boston, Mass.: Harvard Management Communication Letter (5)5:4.

3. Clay T. 1997. *Odwalla, Inc.* Boston. Mass.: Harvard Business School Publishing Corporation. 33 p.

4. Doeg C. 1999. *Crisis Management in the Food and Drinks Industry: A Practical Approach.* Norwell, Mass.: Kluwer Academic Publishers. 250 p.

5. Fryer B. 2001. *When No New is Good News.* Boston, Mass.: Harvard Business School Publishing Corporation. 10 p.

6. *Investigations Operations Manual.* 2004. Washington, D.C.: U.S. Food and Drug Administration.

7. Klein N, Greyser SA. *The Perrier Recall: A Source of Trouble.* Boston, Mass.: Harvard Business School Publishing Corporation. 4 p.

8. Smith NC, Thomas RJ, Quelch JA. 1996. *A Strategic Approach to Managing Product Recalls.* Boston, Mass.: Harvard Business School Publishing Corporation. 12 p.

9. Smith NC. 1992. *Black & Decker Corporation: Spacemaker Plus Coffeemaker (A).* Boston, Mass.: Harvard Business School Publishing Corporation. 12 p.

10. Stauffer JE. 1988. *Quality Assurance of Food: Ingredients, Processing and Distribution.* Trumbull, Conn.: Food & Nutrition Press, Inc. 304 p.

Index

Printed in the United States
144742LV00005BA/6/P

9 780813 816395